Forensic Gait Analysis

Forensic Gait Analysis
Principles and Practice

Edited by
Ivan Birch, Michael Nirenberg,
Wesley Vernon OBE and Maria Birch

CRC Press is an imprint of the
Taylor & Francis Group, an **informa** business

CRC Press
Taylor & Francis Group
6000 Broken Sound Parkway NW, Suite 300
Boca Raton, FL 33487-2742

and by CRC Press
2 Park Square, Milton Park, Abingdon, Oxon, OX14 4RN

© 2020 by Taylor & Francis Group, LLC

CRC Press is an imprint of Taylor & Francis Group, an Informa business
No claim to original U.S. Government works

International Standard Book Number-13: 978-1-138-38684-6 (Hardback)
International Standard Book Number-13: 978-0-429-42658-2 (eBook)

This book contains information obtained from authentic and highly regarded sources. Reasonable efforts have been made to publish reliable data and information, but the author and publisher cannot assume responsibility for the validity of all materials or the consequences of their use. The authors and publishers have attempted to trace the copyright holders of all material reproduced in this publication and apologize to copyright holders if permission to publish in this form has not been obtained. If any copyright material has not been acknowledged please write and let us know so we may rectify in any future reprint.

Except as permitted under U.S. Copyright Law, no part of this book may be reprinted, reproduced, transmitted, or utilized in any form by any electronic, mechanical, or other means, now known or hereafter invented, including photocopying, microfilming, and recording, or in any information storage or retrieval system, without written permission from the publishers.

For permission to photocopy or use material electronically from this work, please access www.copyright.com (http://www.copyright.com/) or contact the Copyright Clearance Center, Inc. (CCC), 222 Rosewood Drive, Danvers, MA 01923, 978-750-8400. CCC is a not-for-profit organization that provides licenses and registration for a variety of users. For organizations that have been granted a photocopy license by the CCC, a separate system of payment has been arranged.

Trademark Notice: Product or corporate names may be trademarks or registered trademarks, and are used only for identification and explanation without intent to infringe.

Visit the Taylor & Francis Web site at
http://www.taylorandfrancis.com

and the CRC Press Web site at
http://www.crcpress.com

Contents

Foreword vii
Preface ix
Contributors xi

Chapter 1 Introduction 1
Ivan Birch and Wesley Vernon

Chapter 2 The history of the use of gait analysis as evidence 19
Michael Nirenberg

Chapter 3 Fundamentals of human gait and gait analysis 39
Ambreen Chohan, Jim Richards and David Levine

Chapter 4 The legal context of forensic gait analysis 59

Part 1: The legal context in North America 59
Emma Cunliffe

Part 2: The legal context in the United Kingdom 67
Graham Jackson

Chapter 5 The development of the forensic gait analysis quality
assurance process in the UK 71
Sarah Reel

Chapter 6	Initial contact, preliminary assessment of footage and defining the task	87
	Ivan Birch	
Chapter 7	Analysing the questioned and reference footage	107
	Ivan Birch	
Chapter 8	Comparison of gait and evaluation	117
	Ivan Birch	
Chapter 9	Writing expert witness reports	129
	Part 1: General principles, requirements and pitfalls	129
	Roger Robson and Claire Gwinnett	
	Part 2: Writing forensic gait analysis reports	144
	Ivan Birch	
Chapter 10	Presenting gait evidence in court	147
	Ivan Birch	
Chapter 11	Psychology of perceptual error in forensic practice	159
	Liam Satchell	
Chapter 12	Probative value of gait analysis	171
	Graham Jackson and Ivan Birch	
Chapter 13	Case studies	189
	Case study 1: Berry's One Stop Store robbery	189
	Michael Nirenberg	
	Case study 2: Arson of a business premises	198
	Maria Birch	
	Case study 3: Two armed bank robberies	201
	Ivan Birch	

Appendix 1: The Sheffield features of gait tool 205

Index 213

Foreword

I remember very little about the career advice I received in school. However, I do remember being told that you should try and find something that you love so work doesn't become a chore, and something that will give you the chance to learn for the rest of your life. At school in the late 1970s and early 1980s I recall news reports about the "umbrella murder", the "dingo baby", and the "Yorkshire Ripper". I started to become interested in forensic science and the rest, as they say, is history. My masters' studies in the early 1990s were the first specifically relating to the forensic sciences. Realistically, the basic principles remain unchanged, but advances in science, technology, and understanding of the multidisciplinary nature of the subject matter as a whole mean that there are now a plethora of disciplines recognised as being involved in the scientific investigation of crime.

I was largely unaware of the developments occurring in forensic gait analysis until some ten years ago, even though I had spent time working in the related area of forensic footwear comparison, specifically in relation to its value as an intelligence tool. At that time the pioneers in the field approached the professional body representing the forensic sciences (then named Forensic Science Society, renamed in 2012 as the Chartered Society of Forensic Sciences after the granting of a Royal Charter in recognition of its professionalism) to seek ways of gaining formal recognition of their unique forensic skill set.

During the last 20 years working across the sector it has been fascinating seeing newer disciplines, like forensic gait analysis, evolve. There is pattern in many of the niche areas involving battles for acceptance and agreement resulting in evolution. Getting to the point where a book of this nature is essential for wider understanding is indeed an achievement. I have had the pleasure of working with a number of authors on a range of projects and they have certainly battled adversity to get to this point. It is due to their exceptional knowledge, enthusiasm, professionalism, and tenacity that this point has been reached.

This book addresses the areas of forensic gait analysis that require substantial development. It provides a systematic understanding of the relevant science that underpins gait analysis and more critically how and when the science should be applied in the forensic context. In the UK, the forensic sciences are moving towards much needed regulation of all science used across the criminal justice system. This requires validated and standardised methodologies for the analysis, comparison, and reporting of findings. The extensive use of CCTV footage in forensic gait analysis requires a truly multidisciplinary approach and understanding, and its continual growth, as demonstrated by this substantial piece of work, means that it will stand alone and will withstand the scrutiny of accreditation and regulation. This book aims to be a guide to professional practice and to become the standard text for use in education and training.

I have been lucky to find a truly varied career that is not a chore, and engagement with new and emerging disciplines like forensic gait analysis has taught me, immeasurably, that anything that is worthwhile is worth fighting for. This publication is truly a lasting testimony to that.

Dr Anya L. Hunt
CEO, Chartered Society of Forensic Sciences

Preface

Ivan Birch, Michael Nirenberg, Wesley Vernon and Maria Birch

This book was born out of a combination of necessity and passion. Necessity in that there needed to be an accessible textbook laying out the basics of the use of gait as evidence, and the passion of the editors for this particular area of forensic science practice. The idea of the book is that it will provide a guide to those who wish to develop their skills in forensic gait analysis, and a reference source for all those who are likely to come across forensic gait analysis during the course of their work. The book has been written by authors drawn from a range of scientific backgrounds, the intention being to provide the reader with knowledge and a range of experience informed by practice. The authors are also drawn from a number of different countries, so although the book is predominantly UK and US centric we hope that the content will help to inform your professional practice wherever you are based in the world.

With the recent publication and implementation of the *Code of practice for forensic gait analysis* by the UK's Office of the Forensic Science Regulator, there are now approved professional standards and a quality assurance framework on a par with other areas of forensic science practice. Forensic gait analysis has come of age. It is therefore appropriate that this textbook, wholly dedicated to professional practice, be published at this time. The use of forensic gait analysis continues to increase, as does the number of peer-reviewed research papers considering various aspects of theory and practice. This continued development is due to the dedication of many of those involved in the use of forensic gait analysis, and the help and support of the Office of the Forensic Science Regulator and the Chartered Society of Forensic Sciences.

The chapters are written in a variety of styles of the authors' choosing and include both referenced information and professional opinions. For this we make no apology; it adds depth to the discussion and offers a broader scope of multidisciplinary analysis. The editors would like to thank all the authors for their contributions, without which there would clearly be no book. What the book is not, is a course in forensic gait analysis, although we would hope that it becomes a standard text for all forensic gait analysis courses. We hope that the book will provide you with an introduction to this area of forensic science, but for those considering professional practice it will not teach you everything you need to know to be a competent practitioner. It is your responsibility to find appropriate further reading, courses and conferences, and build on the knowledge and experience shared by the contributors and editors.

Contributors

Professor Ivan Birch is Consultant Expert Witness in forensic gait analysis with Sheffield Teaching Hospitals NHS Foundation Trust and Emeritus Professor of Human Sciences. He graduated in 1978 with a BSc Joint Honours in Science from the University of Salford, UK, earned an MSc in Human Biology from the University of Loughborough, UK, in 1980, and was awarded a PhD in Biomechanics by the University of Brighton, UK, in 2007. He has extensive experience of teaching biomechanics, anatomy, physiology and research methods, and is a Member of the Chartered Society of Forensic Sciences. In 2015 he was awarded the status of Chartered Scientist by the Science Council for his work in gait analysis as evidence. Ivan is included on the National Crime Agency Specialist Operations Centre Expert Witness Advisers Database in the UK, and has 40 years' experience of gait analysis.

Dr Maria Birch is Principal Lecturer and Deputy Head of the School of Health Sciences at the University of Brighton, UK. She has been teaching students of the allied professions for 24 years, mainly lower limb anatomy and clinical podiatry. She has published work in forensic practice, arterial disease and reflective practice. She has presented at national and international conferences on the student experience of learning anatomy, of the teaching and assessment of anatomy and of the promotion of autonomous learning. She supports student research, and peer reviews for several journals. She is a Senior Fellow of the Higher Education Academy, a Fellow of Podiatric Medicine with the College of Podiatry and is registered with the Health and Care Professions Council.

Dr Ambreen Chohan is Research Fellow at the University of Central Lancashire, UK. She has been working and conducting research in biomechanics and related areas for approximately 16 years, completing her PhD in human motor control in 2008, following her initial degree in Biomedical Engineering. She has a growing track-record in research with over 50 peer-reviewed articles, abstracts and conference papers, with the majority involving independently funded collaborative work with small, medium and large companies' product testing. Her other roles include reviewing for around 20 peer-reviewed journals, and she supports undergraduate modules and postgraduate research students in the area of biomechanics and allied health research.

Dr Emma Cunliffe is Associate Professor at the Allard School of Law, University of British Columbia, Canada. Her research considers the reliability of expert forensic evidence in criminal trials, and implicit and structural bias. She is also a member of the Evidence-Based Forensic Initiative, which is based at the University of New South Wales, Australia. Her present research project considers whether expert evidence and other forms of specialist knowledge operate as a Trojan horse by which discriminatory knowledge and effects are reinforced within the legal system. It also addresses the adequacy of legal mechanisms for supervising the reliability of expert claims.

Dr Claire Gwinnett is Professor in Forensic and Environmental Science at Staffordshire University, UK. Her areas of expertise are trace evidence analysis, quality assurance in forensic science and creation of forensic databases. Her research, particularly in forensic fibre examinations and microplastics, has gained international recognition and has been shown to impact industry, the public sector, academia and policy makers. She has generated extensive national and international collaborations, is part of research bodies and networks that influence policy and practice globally and conducts wide-ranging activities to engage the public in her research areas. She has conducted casework in trace particulate analysis for a variety of cases, including hair analysis for wildlife crime investigations. She acts as an expert advisor in the development of validation studies for ISO accreditation for UK police forces. She has worked closely with National Police Chiefs Council working groups for three years, providing expert opinion and professional reviews. She has also developed practitioner workshops for all police forces.

Professor Graham Jackson is a Consultant Forensic Scientist specialising in forensic inference and the communication of expert evidence. He was an Operational Forensic Scientist with the Metropolitan Police Forensic Science

Laboratory and the UK Home Office's Forensic Science Service (FSS) between 1971 and 2004. His operational role covered all stages from scene examination through to laboratory analysis, report-writing and the presentation of expert evidence in court. Other roles included team-leadership, training and development, liaison with clients and stakeholders and managing organisational change. He is a committee member of the Statistics and Law Section of the Royal Statistical Society and a Fellow of the Chartered Society of Forensic Sciences. He has authored or co-authored approximately 30 articles on interpretation and communication of scientific evidence. He is a reviewer for the journals *Science and Justice* and *Forensic Science International*. He is affiliated with Abertay University, UK, as Emeritus Professor of Forensic Science.

Professor David Levine is Professor and the Walter M Cline Chair of Excellence in Physical Therapy at the University of Tennessee at Chattanooga, USA. In addition, he is board certified as a Specialist in Orthopaedics by the American Board of Physical Therapy Specialties and is also certified in dry needling. He has been working and conducting research in many fields including gait analysis since the early 1990s. He continues to practice in physical therapy in addition to his university position. He has presented at over 100 conferences, and has lectured in more than a dozen countries. He has been published in numerous peer-reviewed journals and conference proceedings with over 100 publications. His latest research focusses on control of clinical infectious disease as a founding member of the Clinical Infectious Disease Control Unit at UTC, chronic pain, animal-assisted therapy and laser to improve muscle endurance.

Dr Michael Nirenberg is a Clinical and Forensic Podiatrist. He has assisted in criminal matters involving footprints, footwear and gait, and his analysis has contributed to convictions. He has published on forensic podiatry and clinical-related matters, participated in research and lectured to forensic scientists, law enforcement personnel and physicians. His paper, 'Forensic Methods and the Podiatric Physician', published in 1989, was among the first articles to recognise forensic podiatry as a distinct discipline and was bestowed the Distinguished Podiatric Medical Writing Award. He currently serves as President of the American Society of Forensic Podiatry. He practices podiatric medicine in Crown Point, Indiana.

Dr Sarah Reel is Senior Lecturer in Podiatry at the University of Huddersfield, UK, and a Forensic Podiatrist and Expert Witness with Sheffield Teaching Hospitals NHS Foundation Trust. She is a member of the Forensic Podiatry

Special Advisory Group of the College of Podiatry (CoP), a Fellow of the Royal College of Physicians and Surgeons Faculty of Podiatric Medicine (Glasgow), a Fellow of the Higher Education Academy, a Fellow of the College of Podiatry, an Examiner for the Faculty of Forensic & Legal Medicine of the Royal College of Physicians (London) and also a Chartered Scientist (CoP). She holds a Certificate of Professional Competency in footprint evidence with the Chartered Society of Forensic Sciences and is a registered Expert Witness with the UK National Crime Agency database. She sits on the International Association for Identification General Forensics Committee and is Chair of the Forensic Gait Analysis Quality Standards Writing Group.

Professor Jim Richards is Professor of Biomechanics and Research Lead for the Allied Health Research Unit at the University of Central Lancashire, UK, where he was appointed in 2004. His work includes the clinical application of biomechanics, the development of new assessment tools for chronic disease, conservative and surgical management of orthopaedic and neurological conditions and the development of evidence-based approaches for improving clinical management and rehabilitation. The focus of his work is to encourage inter-professional research and to develop direct parallels with research to the 'real world' of allied health. He has authored over 160 peer-reviewed journal papers and is Associate Editor for *The Knee*. In addition, he is an executive board member of the International Society of Biomechanics Motor Control Working Group.

Roger Robson is President of the Chartered Society of Forensic Sciences, UK. He is a well-respected figurehead and Expert Witness in the forensic industry, specialising in the forensic aspects of serious crimes, especially those involving the transfer of microscopic particles. He has extensive knowledge across most forensic disciplines, their safe application through accreditation and evaluation of the findings in the context of the case. With 40 years' experience, he continues to work on the most complex international criminal cases and advises the UK judicial system, the Forensic Science Regulator, Law Commission and Criminal Case Review Commission on all matters of forensic science and policy formulation.

Dr Liam Satchell is Lecturer in Psychology at the Centre for Forensic and Investigative Psychology at the University of Winchester, UK. His research focusses on 'applied psychology', including topics in forensics and policing. He has written on improving theory and methodology in academic research to better address practitioners' problems and to make psychological science look more like everyday life. This means understanding the psychology of how individuals make sense of the world and other people. He has previously published academic research on the role of gait in first impressions and how biomechanics can communicate aggressive tendencies.

Professor Wesley Vernon OBE is Visiting Professor at the University of Huddersfield, UK. Now retired, he was previously Head of Podiatry, Research Lead and Deputy AHP Lead at Sheffield Primary and Community Services, and Visiting Professor at Staffordshire University, UK. His past roles include Chair of Quality Standards at the Chartered Society of Forensic Sciences (CSFS), Chair of the Forensic Podiatry Science and Practice Committee at the International Association for Identification (IAI) and President of the Society of Shoefitters. He is a Fellow of the CSFS, a Dedication to Service Award holder and Distinguished Member of the IAI and a Meritorious Award holder and Fellow in Podiatric Medicine at the College of Podiatry (CoP). He is a Founding Fellow of the Faculty of Podiatric Medicine at the Royal College of Physicians and Surgeons of Glasgow, UK. He is co-author of the first and second editions of *Forensic Podiatry: Principles and Methods*, and has authored or co-authored over 80 journal articles, textbook chapters and editorials. He has been awarded an OBE (Officer of the Order of the British Empire) for his services to medicine and healthcare.

Me, I'm just a lawnmower,
You can tell me by the way I walk
Genesis, *Selling England by the Pound*, 1973

Introduction

Ivan Birch and Wesley Vernon

G ait is the manner or style in which we undertake a locomotor activity, such as walking or running (Levine, Richards, and Whittle 2012). 'Gait' and 'walking' are therefore not synonymous and should not be used interchangeably. The process of walking is a learnt motor pattern, a complex series of motor commands and sequencing, developed as we learn to walk, stored in the brain and honed by years of experience. Our gait is the product of a combination of this developmental process that accommodates our anatomy, physiology, injury and pathology as intrinsic factors, and the effects of variables such as terrain, clothing including footwear and obstacles as extrinsic factors. The process of gait is the combined responsibility of several areas of the brain, and is largely subconscious, the advantage of which is the speed and fluidity with which the process is executed.

Gait analysis is the systematic study of human walking, using the eye and brain of experienced observers, which can be augmented by instrumentation for measuring body movements, body mechanics and the activity of the muscles (Levine, Richards, and Whittle 2012). The first known written reference to gait analysis was made by Aristotle (384–322 BC), who observed that if a reed dipped in ink was attached to the head of a man who walked alongside a wall, the line produced would not be straight, but would deviate up and down (Baker 2007). Unfortunately, as noted by Baker (Peck and

Forster 1968, Baker 2007), at the time of this observation it was generally believed that scientific truth could be determined by thinking about a problem, and as a result, while this basic observation is true, the majority of the related propositions are now known to be false. With the emergence and development of science and mathematics during the Renaissance, the underpinning concepts of modern gait analysis began to be developed by workers such as Cardan (1501–1576) and Descartes (1596–1650), who developed methods of describing the orientation of a rigid body in space, and perhaps most significantly Borelli (1608–1679), who carried out research into the contractile properties of muscle and locomotion (Baker 2007). With the development of Newtonian mechanics and its application to human movement, workers such as Boerhaave (1668–1738), Wilhelm Eduard Weber (1804–1891), Ernst Heinrich Weber (1795–1878), Eduard Friedrich Wilhelm Weber (1806–1871) and Guillaume Duchenne (1806–1875) established the science that underpins gait analysis. The development of the science and understanding has continued unabated, drawing on the expertise of a range of professions and finding new applications, driven by workers such as Marey (1830–1904), Carlet (1849–1892), Braune (1831–1892), Fischer (1861–1917), Bernstein (1896–1966), Inman (1905–1980), Murray (1925–1984) and Perry (1918–2013) (Sutherland 2001, Baker 2007, Al-Zahrani and Bakheit 2008).

Gait analysis is used extensively in the diagnosis, treatment planning and evaluation of interventions in many areas of healthcare and rehabilitation, and is also widely used in sports science. High impact peer-reviewed journals[1] and international conferences regularly publish and disseminate research and case studies on all aspects of gait analysis. As a result of this long-standing developmental process, gait analysis has become established as a widely used science.

Forensic gait analysis is the analysis, comparison and evaluation of features of gait to assist the investigation of crime (Code of practice for forensic gait analysis 2019[2]). This is the most recent definition and it captures the nature of the work in appropriate and accurate language. The initial description of forensic gait analysis, and the one that has until recently been the most widely referenced, was that of Kelly in 2000. This is usually referenced as being that of Buncombe (Buncombe 2000), although this reference is actually a newspaper article reporting on a case at the Central Criminal Court in London (the Old Bailey) at which Kelly was giving evidence. Kelly described forensic gait analysis as being "the identification of a person or persons by their gait or features of their gait". This definition served the area of work well for a number of years. However, use of the word 'identification' is problematic. Forensic gait analysis as currently most widely practised is based on observational gait analysis, in which features of gait are observed and noted. A feature of gait is "A kinematic attribute of the gait of a person that can be seen in video footage. Features of gait include angular relationships, segmental orientations and temporal and spatial displacements" (*Code of practice for forensic gait analysis* 2019, p.43). Features of gait are what can be described as class characteristics, which are those features which show consistency and compatibility with a group or proportion of a population, but are not unique (Tuthill and George 2002, Gardner and Krouskup 2016). As such, while features of gait, and in particular combinations of features of gait, can be highly discriminatory, they are not individual characteristics, limited to a single person (Figure 1.1).

In most cases the use of such features cannot result in the identification of a person from the general population, although in particular circumstances where a closed or limited population is being considered, the identification of a single person from a very limited range of possibilities may be possible. The use of 'identification' in the Kelly description could also lead to the misconception that the objective of forensic gait analysis is to achieve a positive identification, which it is not. The actual objective is an assessment of the compatibility or lack of compatibility of observed features of gait. This issue was addressed by DiMaggio and Vernon (2011) whose definition was "the recognition and comparison of gait and features of gait, to assist in the process of identification". The purpose of the 2019 definition, which forms the basis of the *Code of practice for forensic gait analysis*, was to make clear that there is a particular process involved (analysis, comparison and evaluation, the content of the document making it clear that there is also an additional final stage, that of independent verification), as would be expected of any forensic discipline.

FIGURE 1.1 Walking is a complex process that we perhaps take for granted, comprised of numerous class level features of gait, the combination of which can contribute to identification.

4 Chapter 1. Introduction

As the Kelly description, reported by Buncombe (2000), was widely used during the development of forensic gait analysis post 2000, it has resulted in an unforeseen misapprehension. The reporting of the case by Buncombe in 2000, in which for the first time an expert witness compared the gait of a perpetrator captured on video to that of a suspect captured on video, led to the assertion that the use of gait as evidence was a new concept and therefore that gait analysis was also a new concept. As we have discussed, gait analysis has a history going back more than 2000 years, and the use of gait as evidence can be traced back to 1839 (Trial of Thomas Jackson 1839). Interestingly, the 1839 case and the 2000 case were both heard at the same court, the Old Bailey, London, England (Figure 1.2).

Why has the use of forensic gait analysis increased so dramatically in the last 20 years? Cases in the UK that have involved gait analysis now number in the hundreds and its use in the rest of Europe, Canada and the USA has also been reported. The reason for this increase in use is multifactorial.

Although gait analysis in a forensic context has been utilised for many years (Nirenberg, Vernon, and Birch 2018), it is only in recent times that this has been widely publicised. The reporting of the 2000 case at the Old Bailey in London (Buncombe 2000) resulted in the listing of the case by Guinness World Records as "The first time that forensic gait analysis (the analysis of a person's style of walking as a method of identification) has been admissible

FIGURE 1.2 The Central Criminal Court of England and Wales, usually referred to as the Old Bailey after the street in which it is located, was the location of both the first known use of gait analysis as evidence in a trial in 1839, and perhaps the most well known using an expert witness, in 2000.

as evidence in criminal law", raising awareness in the mass media (Guinness World Records 2018). A writer of crime fiction has included a podiatrist specialising in forensic podiatry and gait analysis in his novels, again raising awareness of forensic gait analysis in the general public. From a forensic science perspective, of much greater importance has been the growing body of research activity and publication. Research into the use of gait as a contributor to identification dates back to the work of Cutting and Kozlowski (1977), and has become progressively more focussed, dealing with specific aspects and challenges that the concept presents (Hayfron-Acquah, Nixon, and Carter 2003, Hossain et al. 2010, Bouchrika et al. 2011, Iwashita, Stoica, and Kurazume 2012, Birch et al. 2013a, Choudhury and Tjahjadi 2013, Birch et al. 2014, Lynnerup and Larsen 2014, Birch, Birch, and Bray 2016, Nirenberg, Vernon, and Birch 2018). The research has also demonstrated a number of variant strands to the use of gait in the forensic context, which will be considered later in this text. There has been a marked increase in the number of publications and presentations on forensic gait analysis in scientific journals and at professional conferences. This has contributed to raising awareness within the forensic science and criminal justice communities, which in turn has led investigators to consider forensic gait analysis when faced with the task of attempting to identify perpetrators of crime captured on CCTV recordings. Increasing exposure of the concept of gait analysis providing evidence to the general public, the scientific community and those working in the criminal justice system is a starting point, but perhaps a much more significant factor in the acceleration of growth of use has been casework itself. None of the awareness raising described above would lead to a sustained increase in the use of forensic gait analysis if it had not proved to be a genuinely useful contributor to the casework being undertaken by the commissioning agency. When coupled with a careful and diligent approach on the part of those involved, the result is positive feedback from the commissioning agencies and perhaps most importantly word of mouth recommendation to other members of the legal professions and criminal justice community. But what can forensic gait analysis provide that can complement other forensic sciences?

Gait analysis can be undertaken on footage that does not offer enough detail for the application of other forensic strategies such as facial recognition. While perpetrators of crime can obscure their face and keep their backs to cameras, they still have to get to and from the scene of a crime, which is often achieved by walking. Perpetrators of crime are gaining awareness of mainstream forensic strategies that can contribute to identification, and modify their behaviour accordingly (Smith and Bond 2014). Could this learning lead to them obscuring the way they walk? It is relatively easy to conceal skin colour or the face. Obscuring or changing the way they walk is more complex. Obscuring the way you walk from view by a camera is perhaps the easier option. A question often asked of practitioners giving gait evidence in court is can clothing affect gait? The answer is yes it can, but with the exception of footwear, only in extreme circumstances, such as the wearing of long, slim-fitting pencil skirts. What clothing can do more easily is obscure the way someone walks from view by a camera by making the outline of the body less easily identifiable. Intentionally bringing about an actual change in the way someone walks, without producing a ludicrous result, while not impossible, is more complex and difficult to sustain. As we have said, walking is a learnt motor pattern, and conscious change of the way we walk requires the intervention of other areas of the brain in this well-rehearsed process, reducing the speed and fluidity with which the process can be executed. Gait can therefore be consciously changed, but to change one's gait in the short term to a gait that appears natural and not obviously amended is not easily achievable. Another consideration is that while some features of gait are the result of habitual patterns of motor control, some are the result of anatomical structure such as the orientation of the knee relative to the orientation of the foot. While the neurological control of locomotion can be altered consciously, the anatomy cannot. So while some features of gait can be altered by thinking about them, others cannot. Gait, and particularly what might be called usual gait, the habitual gait we revert to when there are no significant extrinsic factors in play, therefore has good potential as a contributor to identification. Gait can be seen to varying degrees in CCTV footage of relatively low quality. It is difficult to alter convincingly and its use as evidence has developed a good case-based track record in assisting both prosecution and defence in establishing the truth.

Despite the increase in use of gait as evidence, there are challenges that need to be addressed. Observational gait analysis can at best be a contributor to identification, and this needs to be clearly understood by all those associated with its use. One of the problems facing the use of gait analysis as evidence is the misplaced assumption that the goal is identification. Forensic gait analysis based on observational data, and as described and discussed in this book, does not seek to achieve a goal of identification, simply to offer data that can contribute to a wider process of identification or exclusion. A goal of achieving identification based solely on gait analysis would be based on the assumption that gait is unique. Is habitual gait unique? The answer is that at the time of writing we do not know, and in view of the consideration of the concept of uniqueness by Page et al. in 2011 we may never know (Page, Taylor, and Blenkin 2011). If we consider the complexity of the process, the potential variability of its control, and the anatomy and physiology of its execution, it seems highly likely that gait is at least highly discriminatory as an identifier. However, to detect the subtle variations such factors may have on gait requires a considerable amount of data, data that could easily be obtained in a gait laboratory, but is far beyond that available from uncontrolled video footage alone. The considerable amount of research that is currently being undertaken into the development of gait as a biometric suggests that there is a general belief that gait is at least highly discriminatory as an identifier, if not unique, but as noted earlier, currently we do not have enough data on which to base such an assumption. Identification is therefore never the objective of forensic gait analysis as currently practised. The 2017 *Forensic gait analysis: a primer for courts*, published by The Royal Society and the Royal Society of Edinburgh, appears to have been based on the misapprehension that the goal of forensic gait analysis is identification (The Royal Society and the Royal Society of Edinburgh 2017). While some may consider identification from gait to be an achievable aspiration, it has no relevance to current or currently foreseeable forensic gait analysis practice.

Forensic gait analysis has proved that it has genuine value in forensic casework, but its utility alone is not a suitable platform for continued use and development. While gait analysis has a solid and long-standing research evidence base (Baker 2007, Wren et al. 2011), the research evidence base for its use in the forensic context (Birch et al. 2013a, Birch et al. 2013b, Birch et al. 2014, Birch, Birch, and Bray 2016, Birch, Gwinnett, and Walker 2016) is still growing to meet the increasing demands of forensic science practice. A great deal of time and effort is being invested in undertaking such research and expanding our knowledge and understanding of the strengths and weaknesses of using gait as evidence. The evidence base for forensic gait analysis has been steadily increasing over the past 10 years and this has contributed to the increasing use of the speciality in the criminal justice system. Of particular relevance to this book is the work the authors and their colleagues have been undertaking that has culminated in the development and testing of the Sheffield Features of Gait Tool (Birch et al. 2019) (see Appendix 1). This is the first such tool to be developed and offers the first forensic gait analysis strategy to be tested for repeatability, reproducibility and validity (Birch, Birch, and Otten 2018). This is a timely development in view of the production of the *Code of practice for forensic gait analysis* (2019).

Professional practice in forensic gait analysis must be based on the use of "validated methods or procedures based on sound scientific principles and methodology" (Forensic Science Regulator 2017, p.16). The use of observational gait analysis is interpretive, and where interpretive approaches are utilised in forensic analysis practitioners must be able to "demonstrate that they can provide consistent, reproducible, valid and reliable results that are compatible with the results of other competent staff" (Forensic Science Regulator 2017, p.36). In the US, the Supreme Court established a list of factors, usually referred to as 'Daubert factors', that trial judges should consider when reaching a decision on whether scientific evidence derives from scientific knowledge (*Daubert v. Merrell Dow Pharmaceuticals* 1993, Moenssens, DesPortes, and Edwards 2007, Giannelli 2009, Vanderkolk 2009, Pyrek 2010):

- ✦ testing and testability
- ✦ peer review and publication
- ✦ error rate
- ✦ maintenance of standards
- ✦ general acceptance

The UK Forensic Science Regulator's *Codes of practice and conduct* (2017) and the Daubert criteria, which clarify Fed. R. Evid. 702, provide a clear indication as to how professional practice, the underpinning science and the evidence base associated with the use of gait analysis as evidence has been developing and must continue to develop in the forthcoming years. Contributing to this process is an essential responsibility of all practitioners.

Governance has also been a key focus of development. There have been widespread calls for governance to be improved across all areas of forensic science and practice (National Research Council 2009, Adams et al. 2013). As with any forensic science, without demonstrable governance arrangements being in place it is highly unlikely that forensic gait analysis would maintain acceptability in the criminal justice system. A considerable amount of work has been put into governance for forensic gait analysis practice in recent years. Such work has included defining the role and scope of practice of forensic podiatry, including forensic gait analysis, by the International Association for Identification (Vernon et al. 2010), a competency testing scheme run by the UK's Chartered Society of Forensic Sciences, the availability of M-Level postgraduate courses run by the University of Huddersfield in the UK, and the development and use of in-house Standard Operating Procedures developed by the Forensic Podiatry Team based in Sheffield Teaching Hospitals NHS Foundation Trust UK. The *Code of practice for forensic gait analysis* (2019) was written by a writing group of the Chartered Society of Forensic Sciences' Forensic Gait Analysis Working Group in collaboration with the College of Podiatry and the Office of the UK Forensic Science Regulator, and forms part of the Forensic Science Regulator's *Codes of Practice and Conduct for Forensic Science Providers and Practitioners in the Criminal Justice System*. This is an important step forward for forensic gait analysis both in terms of the quality assurance of professional practice and parity with professional standards expected of all other forensic disciplines. The *Code of Practice* will help to ensure a more standardised approach to casework, and has been specifically constructed in a way that will facilitate its adoption, albeit modified appropriately, by regulatory bodies in other countries. It is anticipated that an established and recognised system of governance associated with the implementation of the *Code of practice for forensic gait analysis* (2019) will serve as another facilitator for the increased use of forensic gait analysis.

Professional practice in forensic gait analysis is therefore developing in line with the expectations placed on all forensic disciplines. Criticism is still to be expected, and indeed welcomed as the basis for development and improvement. Cunliffe and Edmond, Chin and Dallen and more recently van Mastrigt et al. have all identified aspects of the use of gait analysis as evidence that warrant scrutiny and consideration (Cunliffe and Edmond 2013, Chin and Dallen 2016, Edmond and Cunliffe 2016, van Mastrigt et al. 2018). Unfortunately, the papers of Cunliffe and Edmond and Chin and Dallen did not consider the current developments in terms of underpinning research and perhaps more importantly professional standards and quality assurance. Nevertheless, on the basis of the information the authors had at hand, a number of their criticisms were justified at the time of their writing and should inform current and future practice developments. To date there is no evidence to suggest that these criticisms have had a negative impact on the number of forensic gait analysis reports being commissioned.

More recent criticism has come in the form of the 2017 publication of the *Forensic gait analysis: a primer for courts* (The Royal Society and the Royal Society of Edinburgh 2017). The intention of this document was to provide information to courts on the use, probative value and cautions that should be exercised when considering forensic gait analysis. Its publication is of course welcomed as a systematic method of providing appropriate information for courts, but has led to a good deal of critical debate. While the document highlights key limitations of the use of gait as evidence, its consideration of the available literature is limited, and as already discussed, it seemingly misinterprets the intent of current practice. Forensic gait analysis compares the features of gait observed to be exhibited by a figure in one set of footage with those exhibited by a figure in another set of footage. Comparison of the two sets of features observed results in one of three possible outcomes: weighted support for one of two opposing propositions, or no support for either of the opposing propositions. Identification is never the outcome, nor is it the objective. The primer also provides a rather simplistic assessment of the probative value that can be placed on forensic gait analysis evidence, based on the findings of a number of research papers (Cutting and Kozlowski 1977, Kozlowski and Cutting 1977, Stevenage, Nixon, and Vince 1999, Troje, Westhoff, and Lavrov 2005, Birch et al. 2013a). At the time of writing, this too is an area of debate, one element of which is again the fact that the research papers involved were investigating the concept of identification from gait rather than its role as a contributor to identification.

Another area of challenge has been that of cost. Forensic gait analysis casework is time-consuming, largely due to the number of hours needed to get footage to play using a fit-for-purpose player (a significant time-consuming task in the experience of the authors), identify the relevant sections showing the figure of interest from the total recordings that have been submitted for consideration and then analyse the gait of the figure of interest. Footage is played multiple times at full speed, in slow motion, forwards and sometimes backwards. In the early days of use of forensic gait analysis, the submission of unsuitable footage for analysis, either in terms of content or quality, also contributed to increased overheads for the analyst, and wasted time for prospective commissioning agencies. While the playability of footage is still a significant issue, the large scale submission of inappropriate footage has to a great extent been resolved as a result of the diligent efforts of forensic gait analysts, who often provide basic protocols written to assist the submitting agencies to help them understand what is needed for a forensic gait analysis to take place optimally and efficiently. Feedback is provided on unsuitable footage, and advice provided to commissioning agencies on what constitutes good footage for the purposes of forensic gait analysis. As the use of forensic gait analysis has increased, commissioning agencies have learnt to identify routinely what is good and bad footage themselves. As a result, the submission of unsuitable footage for analysis has decreased and the process has become more cost effective.

So far we have concentrated on the use of observational gait analysis as the means of collecting gait data for comparison, and indeed the rest of this text will continue to focus on the use of observational gait analysis. The reason for this is that by far the majority of gait analysis that has been utilised in casework has been observational gait analysis. Other methods have been used (Lynnerup and Vedel 2005, Bouchrika et al. 2011), but their use has to date been limited. The use of gait as a contributor to identification can perhaps be divided into three categories: observational gait analysis from video footage, measurements from video footage, and gait as a biometric.

Observational gait analysis from video footage

Drawing on the research and commonly used practices often associated with healthcare, observations are made by eye of the features of gait of a figure seen in questioned footage and a subject seen in reference footage. Comparisons are then made and conclusions drawn based on the knowledge and experience of the analyst. Both the observation of the features of gait and the drawing of conclusions are subjective, and therefore the competency of the analyst is paramount. While the use of purpose-designed software to analyse aspects of gait from video footage can assist the process, and would seem to offer an appropriate area of development for observational gait analysis from video footage as evidence, careful consideration needs to be given to its use. Most such software is designed for use in controlled conditions, where the position and orientation of the subject relative to the camera can be determined prior to video capture, allowing the subject to be filmed largely in one plane with high resolution and good lighting. Its use on footage captured at an oblique angle relative to the subject, often of limited resolution and lighting, introduces challenges beyond the design considerations of the software, that may substantially affect the outcomes of the analysis. Whatever the method by which the features of gait are analysed, a fundamental limitation is currently the lack of data regarding the prevalence of those features in the population. While this does not undermine the determination and comparison of features of gait, it does impinge on the weight that can be placed on subsequent inferences. While projects are ongoing to produce reference databases, such inferences are currently limited to the experience in human gait of those undertaking the work.

A significant misunderstanding that must be addressed is the erroneous definition of forensic gait analysis as being qualitative. Despite the fact that forensic gait analysis using observational gait analysis techniques does not yield measurements, it remains a quantitative technique. Quantitative means "relating to, measuring, or measured by the quantity of something rather than its quality" (*Oxford Dictionary of English*). When features of gait are observed we are observing positions, angles and timings which are quantitative in nature. Flexion of a joint or abduction of a foot has an angle irrespective[3] of our inability to measure that angle. We say that a joint is flexed or extended, or that a foot is abducted or adducted, and in doing so we are referring to a quantity, in this case an angle. The fact that we cannot measure the angle does not make it any less quantitative, and it certainly does not make it qualitative, "relating to, measuring, or measured by the quality of something rather than its quantity" (*Oxford Dictionary of English*). The authors suspect that this misapprehension has inadvertently arisen from a previous publication. However, the fact that it is heard repeated is of some concern as it would suggest that in some cases those repeating the misapprehension either have not considered the meaning of what is being said, or that they simply do not understand. In either case, the authors would hope that any forensic scientist would reflect on such a fundamental aspect of professional practice.

Measurements from video footage

Prior to the advent of three-dimensional motion analysis systems, much biomechanical analysis was based on the frame by frame XY co-ordinate plotting of anatomical landmarks, undertaken by the analyst. As automated two- and three-dimensional motion capture became more widely available, this methodology largely fell into disuse by biomechanics. More recently, work has been undertaken to investigate the use of such strategies as a contributor to identification (Larsen et al. 2010, Yang et al. 2014). Although the strategy renders approximations of distances and joint angles, within the limitations of parallax and virtual marker positioning, it remains a fundamentally subjective process if the positioning of the virtual markers relies on manual placement. The use of cameras designed for computer gaming offers an interesting development in the ability to gain objective and automated three-dimensional information from a single camera, and development of this strategy as a contributor to identification continues (Ball et al. 2012, Preis et al. 2012, Kastaniotis et al. 2015). As is the case with observational gait analysis in the forensic context, the lack of substantive databases regarding the prevalence of features of gait means that any estimations of the significance of findings are often largely reliant on the experience of those undertaking the work or population specific ad hoc surveys.

Gait as a biometric

There has been a considerable investment of resource in the use of gait as a biometric,[3] either as a standalone variable or in combination with some form of body mapping (Huang, Harris, and Nixon 1999, Nixon et al. 2002, Cunado, Nixon, and Carter 2003, Hayfron-Acquah, Nixon, and Carter 2003, Nixon 2008, Iwashita, Stoica, and Kurazume 2012). This body of work is predicated on the notion that gait is unique (Yam, Nixon, and Carter 2004, Kastaniotis et al. 2015, Semwal, Raj, and Nandi 2015). While this may be true, the question is whether enough information can be gained regarding gait from a single piece of video footage to establish uniqueness? This approach to the use of gait in identification offers the possibility of a truly objective methodology, but has yet to be achieved in a way that can be meaningfully and consistently applied in an uncontrolled environment. One area of work where observational gait analysis and biometric analysis overlap is that of soft biometrics, the use of traits that "are physical, behavioural or adhered human characteristics, classifiable in pre-defined human compliant categories" (Dantcheva et al. 2011). Reid and Nixon considered the use of comparative descriptors during gait as a method of identification and their work has similarities to the systematic descriptions of features of gait used in observational gait analysis (Reid, Nixon, and Stevenage 2014).

Forensic science in general, and the more recently developed areas of practice in particular, are constantly being scrutinised and challenged. We live in a world rich with science, a cyclic process of theory, test and revision. Science underpins forensic practice, and has since the ancient Greeks and Romans, Archimedes usually being attributed with the first use of forensic science to demonstrate the theft of gold (Schafer 2008). But while we live in a world rich with science, we do not live in a world rich with the understanding of science. Science provides information to support or refute theories, based on our knowledge as it is at the time of testing, but it does not provide absolute answers. One of the most significant challenges to forensic science is the lack of understanding of the nature of data, what conclusions can be drawn from those data, and what weight can be placed on those conclusions. It is very tempting, or even reassuring, for those with less understanding of the nature of science to assume that science can answer all questions unequivocally.

Forensic science should be questioned and challenged as part of the scientific process. However, the challenges should be predicated on the notion of improvement, assist in the development of the science, and be informed by the understanding that the answers are never absolute. A key aspect of forensic science is that of probative value.[4] How much evidential weight can be placed on a finding or conclusion? As the science of a particular area of forensic practice develops and the theories are tried, tested and revised, it is likely that the probative value of evidence based in that area will increase. The longer the area of practice is developed, the better the data is understood, the more reliable the outcomes. This assumes that the area of practice is genuinely tried and tested in a robust scientific manner. This is the process of science. The development of new areas of forensic practice is therefore inevitably protracted, and should be accompanied by a cautious increase of the probative value of the evidence they produce.

Challenges to new or innovative areas of forensic practice are therefore necessary, but should be informed. Is the concept sound? Does the area of work have a scientific evidence base? What is the probative value of the data produced at this stage in the development of this area of practice? New areas of forensic science should be scrutinised and challenged, as part of the scientific developmental process. Established forensic science disciplines such as fingerprints were subsumed into practice with little or no research underpinning their use, the research and understanding developing along with the area of forensic practice and challenges and criticisms of its use (Zabell 2005, Dror, Charlton, and Péron 2006, Vokey, Tangen, and Cole 2009). This process inevitably involves misuse of the concept during the early stages of its development. The alternative strategy is to research the concept, and not use it until all the information is gained as to its uses and limitations. Of course from a scientific perspective this can never actually be fully achieved. This latter process inevitably results in opportunities for using the concept to help solve a crime being missed. Both strategies are to some degree unethical. The critical factors are that the research is being undertaken, the understanding is being developed, and that an informed approach is taken to the determination of the weight that can be placed on the data gained. Unfortunately, we will still live in a world where some forensic strategies are considered to be capable of unequivocal identification and others are dismissed as having no value.

Forensic gait analysis has come to the fore at a time of challenge to the forensic sciences, a time when the importance of repeatability, reproducibility, reliability and validity are well known, although perhaps not so well understood.

NOTES

1. Examples of such journals include *Gait and Posture*, *Clinical Biomechanics*, the *Journal of Biomechanics* and the *Journal of Applied Biomechanics*.
2. This document was written by The Chartered Society of Forensic Sciences' Forensic Gait Analysis Working Group in collaboration with the College of Podiatry, who were tasked by the Forensic Science Regulator to write a code of practice for forensic gait analysis capable of being read as a self-contained or standalone document. Now that the code of practice has been published, the Forensic Science Regulator requires the provider of forensic gait analysis to ensure their services comply with the requirements outlined.
3. Biometrics, or perhaps more accurately biometric recognition, is the "automated recognition of individuals based on their biological and behavioural characteristics" (ISO/IECDIS2382-37 Information technology – Vocabulary – Part 37: Biometrics 2012).
4. Probative value is the extent to which evidence can be relied upon to demonstrate a fact.

REFERENCES

Adams, Dwight E., John P. Mabry, Mark R. McCoy, and Wayne D. Lord. 2013. "Challenges for forensic science: New demands in today's world." *Australian Journal of Forensic Sciences* 45 (4):347–355.

Al-Zahrani, K. S., and M. O. Bakheit. 2008. "A historical review of gait analysis." *Neurosciences (Riyadh, Saudi Arabia)* 13 (2):105.

Baker, Richard. 2007. "The history of gait analysis before the advent of modern computers." *Gait and Posture* 26 (3):331–342.

Ball, Adrian, David Rye, Fabio Ramos, and Mari Velonaki. 2012. "Unsupervised clustering of people from 'skeleton' data." *Proceedings of the Seventh Annual ACM/IEEE International Conference on Human-Robot Interaction*. Conference held in Boston, Massachusetts: Association for Computing Machinery.

Birch, I., L. Raymond, A. Christou, M. A. Fernando, N. Harrison, and F. Paul. 2013a. "The identification of individuals by observational gait analysis using closed circuit television footage." *Science and Justice* 53 (3):339–342. doi: 10.1016/j.scijus.2013.04.005.

Birch, I., W. Vernon, J. Walker, and J. Saxelby. 2013b. "The development of a tool for assessing the quality of closed circuit camera footage for use in forensic gait analysis." *Journal of Forensic and Legal Medicine* 20 (7):915–917. doi: 10.1016/j.jflm.2013.07.005.

Birch, I., W. Vernon, G. Burrow, and J. Walker. 2014. "The effect of frame rate on the ability of experienced gait analysts to identify characteristics of gait from closed circuit television footage." *Science and Justice* 54 (2):159–163. doi: 10.1016/j.scijus.2013.10.002.

Birch, I., T. Birch, and D. Bray. 2016. "The identification of emotions from gait." *Science and Justice* 56 (5):351–356. doi: 10.1016/j.scijus.2016.05.006.

Birch, Ivan, Claire Gwinnett, and Jeremy Walker. 2016. "Aiding the interpretation of forensic gait analysis: Development of a features of gait database." *Science and Justice* 56 (6):426–430.

Birch, Ivan, Maria Birch, and Bert Otten. 2018. "The development and testing of the Sheffield features of gait tool." *International Association for Identification Annual International Educational Conference*, San Antonio, TX.

Birch, Ivan, Maria Birch, Lucy Rutler, Sarah Brown, Libertad Rodriguez Burgos, Bert Otten, and Mickey Wiedemeijer. 2019. "The repeatability and reproducibility of the Sheffield features of gait tool." *Science and Justice* 59 (5):544–551.

Bouchrika, Imed, Michaela Goffredo, John Carter, and Mark Nixon. 2011. "On using gait in forensic biometrics." *Journal of Forensic Sciences* 56 (4):882–889.

Buncombe, A. 2000. "Gang leader is unmasked by his bandy-legged gait." *The Independent*, July 2000, 115.

Chartered Society of Forensic Sciences and College of Podiatry in association with the Forensic Science Regulator. 2019. *Code of practice for forensic gait analysis*, Issue 1. Birmingham: The Forensic Science Regulator.

Chin, Jason M., and Scott Dallen. 2016. "R. v. Awer and the dangers of science in sheep's clothing." *Criminal Law Quarterly* 64 (4):527–555.

Choudhury, Sruti Das, and Tardi Tjahjadi. 2013. "Gait recognition based on shape and motion analysis of silhouette contours." *Computer Vision and Image Understanding* 117 (12):1770–1785.

Cunado, David, Mark S. Nixon, and John N. Carter. 2003. "Automatic extraction and description of human gait models for recognition purposes." *Computer Vision and Image Understanding* 90 (1):1–41.

Cunliffe, Emma, and Gary Edmond. 2013. "Gaitkeeping in Canada: Missteps in assessing the reliability of expert testimony." *Canadian Bar Review* 92:327–368.

Cutting, J. E., and L. T. Kozlowski. 1977. "Recognizing friends by their walk - gait perception without familiarity cues." *Bulletin of the Psychonomic Society* 9 (5):353–356.

Dantcheva, Antitza, Carmelo Velardo, Angela D'angelo, and Jean-Luc Dugelay. 2011. "Bag of soft biometrics for person identification." *Multimedia Tools and Applications* 51 (2):739–777.

Daubert v. Merrell Dow Pharmaceuticals, Inc., 509 U.S. 579, 113 S. Ct. 2786, 125 L. Ed. 2d 469, 1993.

DiMaggio, John A., and Wesley Vernon. 2011. *Forensic Podiatry: Principles and Methods*. New York/London: Humana.

Dror, Itiel E., David Charlton, and Ailsa E. Péron. 2006. "Contextual information renders experts vulnerable to making erroneous identifications." *Forensic Science International* 156 (1):74–78.

Edmond, Gary, and Emma Cunliffe. 2016. "Cinderella story: The social production of a forensic science." *Journal of Criminal Law and Criminology* 106:219.

Federal Rule of Evidence 702.

Forensic Science Regulator. 2017. *Codes of Practice and Conduct: For Forensic Science Providers and Practitioners in the Criminal Justice System, Issue 4*. Birmingham: The Forensic Science Regulator.

Gardner, Ross M., and Donna Krouskup. 2016. *Practical Crime Scene Processing and Investigation*. Second ed. Boca Raton, FL: CRC Press. Original edition, 2012.

Giannelli, Paul C. 2009. "Daubert "Factors"." *Criminal Justice* 23 (4):42–44.

Guinness World Records. 2018. "First use of forensic gait analysis evidence in court." *Guinness World Records*, accessed October 19 2018. www.guinnessworldrecords.com/world-records/first-use-of-forensic-gait-analysis-evidence-in-court.

Hayfron-Acquah, James B., Mark S. Nixon, and John N. Carter. 2003. "Automatic gait recognition by symmetry analysis." *Pattern Recognition Letters* 24 (13):2175–2183.

Hossain, Md Altab, Yasushi Makihara, Junqiu Wang, and Yasushi Yagi. 2010. "Clothing-invariant gait identification using part-based clothing categorization and adaptive weight control." *Pattern Recognition* 43 (6):2281–2291.

Huang, Ping S., Chris J. Harris, and Mark S. Nixon. 1999. "Recognising humans by gait via parametric canonical space." *Artificial Intelligence in Engineering* 13 (4):359–366.

ISO/IECDIS2382-37 Information Technology-Vocabulary-Part 37: Biometrics. 2012. Organization for Standardization and International Electrotechnical Committee.

Iwashita, Yumi, Adrian Stoica, and Ryo Kurazume. 2012. "Gait identification using shadow biometrics." *Pattern Recognition Letters* 33 (16):2148–2155.

Kastaniotis, Dimitris, Ilias Theodorakopoulos, Christos Theoharatos, George Economou, and Spiros Fotopoulos. 2015. "A framework for gait-based recognition using Kinect." *Pattern Recognition Letters* 68:327–335.

Kozlowski, Lynn T., and James E. Cutting. 1977. "Recognizing the sex of a walker from a dynamic point-light display." *Perception and Psychophysics* 21 (6):575–580.

Larsen, Peter Kastmand, Niels Lynnerup, Marius Henriksen, T. Alkjær, and Erik B. Simonsen. 2010. "Gait recognition using joint moments, joint angles, and segment angles." *Journal of Forensic Biomechanics* 1 (10.4303):2090–2697.1000102.

Levine, David, Jim Richards, and Michael W. Whittle. 2012. *Whittle's Gait Analysis*. Philadelphia, PA: Elsevier Health Sciences.

Lynnerup, N., and J. Vedel. 2005. "Person identification by gait analysis and photogrammetry." *Journal of Forensic Sciences* 50 (1):112–118.

Lynnerup, Niels, and Peter Kastmand Larsen. 2014. "Gait as evidence." *IET Biometrics* 3 (2):47–54.

Moenssens, Andre A., Betty Layne DesPortes, and Carl N. Edwards. 2007. *Scientific Evidence in Civil and Criminal Cases*. New York, NY: Foundation Press.

National Research Council. 2009. *Strengthening Forensic Science in the United States: A Path Forward*. Washington, DC: National Academies Press.

Nirenberg, Michael, Wesley Vernon, and Ivan Birch. 2018. "A review of the historical use and criticisms of gait analysis evidence." *Science and Justice* 58 (4):292–298.

Nixon, Mark. 2008. "Gait biometrics." *Biometric Technology Today* 16 (7):8–9.

Nixon, Mark S., John N. Carter, Jamie D. Shutler, and Michael G. Grant. 2002. "New advances in automatic gait recognition." *Information Security Technical Report* 7 (4):23–35.

Page, Mark, Jane Taylor, and Matt Blenkin. 2011. "Uniqueness in the forensic identification sciences—fact or fiction?" *Forensic Science International* 206 (1–3):12–18.

Peck, Arthur L., and E. S. Forster. 1968. *Aristotle: Parts of Animals, Movement of Animals, Progression of Animals*. Boston: Harvard University Press.

Preis, Johannes, Moritz Kessel, Martin Werner, and Claudia Linnhoff-Popien. 2012. "Gait recognition with kinect." *1st International Workshop on Kinect in Pervasive Computing.* Conference in Newcastle, UK.

Pyrek, Kelly. 2010. *Forensic Science Under Siege: The Challenges of Forensic Laboratories and the Medico-Legal Investigation System*. Philadelphia, PA: Elsevier.

Reid, Daniel A., Mark S. Nixon, and Sarah V. Stevenage. 2014. "Soft biometrics; human identification using comparative descriptions." *IEEE Transactions on Pattern Analysis and Machine Intelligence* 36 (6):1216–1228.

Schafer, E. D. 2008. "Ancient science and forensics." *Forensic Science* 1:41–45.

Semwal, Vijay Bhaskar, Manish Raj, and Gora Chand Nandi. 2015. "Biometric gait identification based on a multilayer perceptron." *Robotics and Autonomous Systems* 65:65–75.

Smith, Lisa, and John Bond. 2014. *Criminal Justice and Forensic Science: A Multidisciplinary Introduction.* London, UK: Macmillan International Higher Education.

Soames, C. 2003. *Oxford Dictionary of English*. Oxford: Oxford University Press.

Stevenage, S. V., M. S. Nixon, and K. Vince. 1999. "Visual analysis of gait as a cue to identity." *Applied Cognitive Psychology* 13 (6):513–526. doi: 10.1002/(sici)1099-0720(199912)13:6<513::aid-acp616>3.0.co;2-8.

Sutherland, David H. 2001. "The evolution of clinical gait analysis part l: Kinesiological EMG." *Gait and Posture* 14 (1):61–70.

The Royal Society and the Royal Society of Edinburgh. 2017. *Forensic Gait Analysis: A Primer for Courts*. London: The Royal Society.

"Trial of Thomas Jackson, alias Johnson, alias Wells (t18391216-382)." 1839. *The Proceedings of the Old Bailey*, accessed July 2019. www.oldbaileyonline.org/

Troje, Nikolaus F., Cord Westhoff, and Mikhail Lavrov. 2005. "Person identification from biological motion: Effects of structural and kinematic cues." *Perception and Psychophysics* 67 (4):667–675. doi: 10.3758/bf03193523.

Tuthill, H., and G. George. 2002. *Individualization: Principles and Procedures in Criminalistics*. Second ed. Jacksonville, FL: Lightning Powder Company.

van Mastrigt, Nina M., Kevin Celie, Arjan L. Mieremet, Arnout C. C. Ruifrok, and Zeno Geradts. 2018. "Critical review of the use and scientific basis of forensic gait analysis." *Forensic Sciences Research* 3 (3):183–193.

Vanderkolk, John. 2009. *Forensic Comparative Science: Qualitative Quantitative Source Determination of Unique Impressions, Images, and Objects.* Cambridge, MA: Academic Press.

Vernon, Wesley, Jeremy Walker, Sarah Reel, Haydn Kelly, Brian Brodie, John DiMaggio, Michael Nirenberg, and Norman Gunn. 2010. "The role and scope of practice of forensic podiatry." *Journal of Foot and Ankle Research* 3 (1):O26.

Vokey, John R., Jason M. Tangen, and Simon A. Cole. 2009. "On the preliminary psychophysics of fingerprint identification." *The Quarterly Journal of Experimental Psychology* 62 (5):1023–1040.

Wren, Tishya A. L., George E. Gorton III, Sylvia Ounpuu, and Carole A. Tucker. 2011. "Efficacy of clinical gait analysis: A systematic review." *Gait and Posture* 34 (2):149–153.

Yam, ChewYean, Mark S. Nixon, and John N. Carter. 2004. "Automated person recognition by walking and running via model-based approaches." *Pattern Recognition* 37 (5):1057–1072.

Yang, S. X., P. K. Larsen, T. Alkjær, E. B. Simonsen, and N. Lynnerup. 2014. "Variability and similarity of gait as evaluated by joint angles: Implications for forensic gait analysis." *Journal of Forensic Sciences* 59 (2):494–504.

Zabell, S. L. 2005. "Fingerprint evidence." *Journal of Law and Policy* 13 (1):143– 177.

2

The history of the use of gait analysis as evidence

Michael Nirenberg

The use of gait analysis as evidence continues to become more accepted by law enforcement, attorneys, and judges as a method of assisting in the identification process. Court decisions, particularly in the United Kingdom and the United States in recent years, show the frequent acceptance of gait evidence and testimony, both from lay witnesses and expert practitioners. Gait evidence has been found to be admissible in numerous criminal trials, and has been upheld on appeal. This chapter will trace the history of gait evidence in judicial proceedings, from its beginnings in the 1800s in England, to its application in more recent cases in the United States.

1839 TO 1959

The first known use of gait as a means of identification in court occurred at the London Central Criminal Court in 1839. Thomas Jackson was accused of returning to the United Kingdom while sentenced to relocation, a felony punishable by death. George Cheney, the Ward Beadle, a type of city constable, was called upon to testify against Jackson. Cheney said that he had custody of Jackson two years earlier when the defendant was charged with burglary. Jackson had a bowed left leg and walked with a limp due to an accident. Cheney testified, "I have not a doubt of his being the man—I know him by his walk" (Thomas Jackson 1839). Cheney's identification of Jackson only by his gait resulted in a conviction (Thomas Jackson 1839).

In the 1800s, criminal suspects were typically identified through eyewitness accounts and photographs; there were no scientific means or records to assist identification. But in 1883, French criminologist and anthropologist Alphonse Bertillon created the first system of physical measurements, photography, and record-keeping that could be used by law enforcement to identify recidivist criminals. His method of measuring body parts to identify criminals was called *signaletics* or *bertillonage* (Bertillon 1896). Bertillon identified measurements of the head and body, the shape formations of the ear, eyebrow, mouth, and eye, as well as unique markings like tattoos and scars, and personality characteristics. In terms of his system for gait analysis, Bertillon found that an individual's gait could vary in multiple ways, such as being slow or fast, or having a short step or a long one. In addition, Bertillon said a person's gait could be light, heavy, tripping, or sedate. Bertillon also detailed types or styles of gait: stiff, measured, gawky, swinging, unsteady, and limping (Bertillon 1896). Bertillon's work reveals an early recognition of the potential for a systematic approach to gait analysis to assist with identification (U.S. National Library of Medicine 2014).

The first known use of gait as evidence in the United States was in Texas in 1908. A criminal case involved gait testimony from a robbery victim, Mr. Van Rooyen, who said that although the men who robbed him concealed their faces, he recognised one of the defendants by the man's voice and "his walk."

A murder in Glasgow, Scotland, in 1908 also resulted in the use of gait testimony (Doyle 1912). The brutal bludgeoning of Marion Gilchrist, an elderly woman, caused a public outcry for expeditious justice. Oscar Slater was her neighbour who had left the country under an assumed name shortly after the incident. When he was found in America, he agreed to return to Scotland, confident that he could prove his innocence. In his New York extradition hearing, testimony was presented by Gilchrist's maid, who never saw the murderer's face but identified Slater due to "some peculiarity in his walk" (Doyle 1912). The maid also told the court in Glasgow about Slater's gait (Doyle 1912).

Both in the United Kingdom and the United States, courts have allowed members of the public to provide observational testimony of the perpetrator's gait. Scottish Police Surgeon Sir Sydney Smith, a forensic pathologist, was the first expert witness to opine on his analysis of an individual's gait in 1937 (Smith 1959). A man was arrested after breaking into a shop and was caught in his stockinged feet. A pair of boots was found, which the cat burglar claimed as his.

In the same district, two other burglaries had occurred a few months prior, and each of the three had similar attributes, including the same means of approach, mode of entry, the approximate time of the break-in, and the intruder's general conduct on the premises. In each instance, the cat burglar left his footwear behind, near where he'd broken in. In one of the cases, a pair of boots had been left, in the other two, it was a pair of shoes (Smith 1959). The man apprehended in the third burglary denied any knowledge of the other two. However, investigators were certain that the same man had committed all three offences. They sent the three pairs of footwear to Smith to ascertain whether they could have been worn by the same man. Smith examined the footwear and based only on the footwear, and without seeing the suspect, he provided a description of the perpetrator's physical description and gait.

"I had still not seen him walk when I told the court in some detail how I thought he walked," Smith said (1959). Smith helped to secure a conviction for the prosecution in all three burglaries. In prison, Smith later examined, photographed, and filmed the convicted thief. He remarked, "We found that the story told by his boots was true in every respect" (1959). This included his gait: "His gait corresponded closely to what I had inferred from the markings on the boots and shoes," Smith wrote (1959).

1960 TO 1999

Most cases concerning gait-related testimony typically involve visual identification of the features of a perpetrator's gait (*State v. Hills* 1961; Appel 1987; *People v. Colbert* 2010; *Commonwealth v. Caruso* 2014). One exception is the case of James Leroy Iverson in which witness testimony focussed on the *sound* of the perpetrator's gait (*State v. Iverson* 1971). A cab driver was accused of strangling two North Dakota women in 1968. The women's bodies were found in one of their apartments after they failed to show up for work. Iverson claimed to be intoxicated and did not remember killing the women (*State v. Iverson* 1971). At his trial, key testimony was given by Robert E. Shepler and his roommate, Bruce Gustafson. Their apartment was directly below the victim's, and they said they could hear sounds coming from the apartment. They were graduate physics students at the University of North Dakota, and both men testified that the victim had several regular male visitors who would come to her apartment very early in the morning

(*State v. Iverson* 1971). Gustafson also noticed one particular visitor by his walk. He testified:

> Well, this sounds funny, but there is one guy... I only heard him once, but they called him the one-legged guy because it was funny the way he walked up and down the stairs. I could tell there was something wrong. The cadence in the steps was different than normal, than a normal person's walk up and down the stairs. One foot hit—when one foot hit, it would sound different than the other foot. Naturally we thought he was lame or had an artificial leg or something like that.
>
> (***State v. Iverson* 1971**)

Iverson had diabetes and had chronic problems with one leg, which required he wear a heavy built-up shoe. He was found guilty by a jury in Grand Forks of murder in the first degree of one woman and in the second degree of the other (*State v. Iverson* 1971).

In 1979, an American case considered shoeprint evidence to determine the perpetrator's gait. Jerry Mark was accused of shooting to death his brother and family in 1976 in their Iowa farm house. Police discovered a trail of shoeprints on the dirt and gravel driveway of the farm. As part of the prosecution's case, it introduced photographs of shoeprints found at the scene. The prints were made by size 11 men's Converse (500) tennis shoes. Investigators never found the shoes that had made the prints (*State v. Mark* 1980). Because the actual shoes worn during the crime were never retrieved, the prosecution was unable to compare the shoes themselves to the prints at the crime scene. Instead, the State concentrated on the physical characteristics of the individual making the prints, and compared those distinctive characteristics to those of the defendant. Two experts, podiatrists Dr. William W. Gronen and Dr. Terry K. Lichty, made the comparisons at trial (*State v. Mark* 1980). Following a biomechanical examination of the defendant, Lichty testified regarding his objective physical findings and measurements of the defendant. Lichty testified that after he completed his physical examination of Mark, he believed he would exhibit the following abnormal or distinctive characteristics when he walked:

+ flat footed
+ toe out, and that it would be greater in the right foot than in the left
+ produce a hard heel strike, which would be more pronounced in the right foot
+ have propulsive toes (*State v. Mark* 1980)

All of his predictions were confirmed when Lichty observed Mark walk. The podiatrist testified that Mark's physical characteristics were all consistent with the characteristics exhibited by the maker of the murder scene shoeprints (*State v. Mark* 1980).

Gronen also examined the shoeprints found at the scene of the murders and subsequently examined the defendant by observing his gait. Gronen testified that all of the characteristics exhibited by the defendant were reflected in the shoeprints found at the scene of the crime. Lichty also testified at trial that there were no inconsistencies between the shoeprints and gait analysis of Mark. Although he was unwilling to state that the defendant, in fact, made the prints found at the scene of the crime, he did opine that the likelihood of two individuals possessing the same combination of distinctive characteristics was "rather low." Mark was convicted (*State v. Mark* 1980). On appeal, the Supreme Court of Iowa found that the podiatrists' analysis placed Mark at the scene of the murders and, as a result, "rendered his identity as the murderer more probable than it would have been without the evidence" (*State v. Mark* 1980). The fact that Mark could not be conclusively identified as the maker of the shoeprints did not make the gait-related testimony inadmissible (*State v. Mark* 1980).

An important development in the use of gait as evidence occurred in 1994, when for the first time the gait evidence relied on the submission of surveillance video footage. In a Pennsylvania robbery trial, the Commonwealth called a witness who had known the robbery suspect for approximately a year and who had observed his "distinctive and easily recognised gait" (*Commonwealth v. Spencer* 1994). She testified that when he walked, he had a "bounce in his step…like rolling off a step" (*Commonwealth v. Spencer* 1994). She was shown the video tape of the robbery, and confirmed that the robber walked like the accused. On appeal, the defendant argued that the witness' testimony was incompetent and should have been disallowed because she wasn't qualified to express an opinion that the robber's gait was similar to his. The Court of Appeals held that her testimony was relevant to establish the identity of the robber. The Court noted:

> Whether testimony constitutes fact or opinion is sometimes a difficult determination to make, for at times it may include elements of both. That appellant walked with a gait similar to that of the robber was such testimony. Whether fact or opinion, however, the testimony was properly received by the trial court.
>
> **(*Commonwealth v. Spencer* 1994)**

The appellate court went on to explain that the Superior Court elected to adopt the approach provided by Rule 701 of the *Federal Rules of Evidence*. This approach is as follows:

> If the witness is not testifying as an expert, his testimony in the form of opinions or inferences is limited to those opinions or inferences which are (a) rationally based on the perception of the witness and (b) helpful to a clear understanding of his testimony or the determination of a fact in issue.
>
> **(*Federal Rule of Evidence 701*)**

The Court of Appeals found that the witness' opinion that the gait of the robber and the accused were similar was rationally based on her perception and was essential to a clear understanding of her testimony. The trial court did not err by allowing the jury to hear her gait testimony. The accused's appeal was dismissed (*Commonwealth v. Spencer* 1994).

In another case two years later, police arrested Lamont Williams for a series of armed robberies at fast-food restaurants throughout Milwaukee County in Wisconsin. Williams was arrested after a victim of one of the robberies saw him at a movie theatre and called the police (*State v. Williams* 1996).

At trial, one of the robbery victims, a former dance instructor, testified that Williams was the robber and that he was certain because of his gait (*State v. Williams* 1996). The witness testified that Williams's way of walking matched that of the robber. The dance instructor said he was attentive to the way people moved. He testified that he had observed the way Williams walked at the preliminary hearing and that he concluded his gait was the same as that of the robber of the restaurant. Further, he testified that the gait of Williams, which was played on video captured at the crime scene, was the same as that of the robber. Williams sought to rebut the victim's eyewitness testimony by introducing the testimony of his mother and sister who he argued would have stated that they were familiar with his gait; that they'd seen the videotape of the robbery; and the robber's gait didn't match his. However, the trial court excluded this testimony as irrelevant, and Williams was convicted (*State v. Williams* 1996). Again, this was eyewitness testimony and a lay person's observation of the video recording of the suspect's gait. Here, the witness was not admitted as an expert, but rather happened to be an individual with heightened awareness of gait because of his background as a dance instructor (*State v. Williams* 1996). In a similar case, a defendant appealed his conviction of attempted murder, claiming that the trial court improperly denied his motion to suppress three pretrial identifications. One of these was from the victim, a clinical psychologist, who testified that in his background and experience, he was trained to focus on a person's facial features, movement, and gait. He testified that this training played a role in his positive identification of the defendant and that he was absolutely certain that the defendant was the perpetrator. The Court of Appeals affirmed the conviction (*State v. Santos* 2007).

A 1998 case involving eyewitness testimony questioned the sufficiency of gait evidence to establish guilt beyond a reasonable doubt (*United States ex rel. Bass v. Ahitow* 1998). The eyewitness in the case identified the accused as the gunman in a line-up, however, his initial identification of the man, with the suspect seated, was tentative. The eyewitness told the police that he was uncertain until he observed the man get up and walk to a table. Then he positively identified the suspect as the gunman, telling the police that his observation of the man's gait "clinched the identification" (*United States ex rel. Bass v. Ahitow* 1998). The officer who testified at trial confirmed this fact, noting that the report he prepared after the line-up reflected the eyewitness' positive identification after having seen the suspect walk. The defendant's appeal was denied (*United States ex rel. Bass v. Ahitow* 1998).

In a 1999 Massachusetts jury trial, the defendant Steven Caruso was convicted of two counts of malicious destruction of property (a car) (*Commonwealth v. Caruso* 2014). Caruso was a regular diner at the restaurant where 32-year-old Sandra Berfield worked as a waitress for 12 years. Caruso would come in every day, sometimes twice a day, and spend hours sitting in Berfield's section, staring at her. She complained that after she rejected Caruso's request for a date, he began stalking her. In 1999, he was convicted of slashing the tires on her car and pouring battery acid in her gas tank. (*Commonwealth v. Caruso* 2014). Caruso filed a motion for a new trial. He claimed that Berfield could not see his facial features and according to her trial testimony, she was only able to recognise Caruso from his distinctive gait (which she had observed at the restaurant). She described the gait as "almost like a bounce as he walks with his arm swinging back and forth, his left arm" (*Commonwealth v. Caruso* 2014). Caruso's motion was denied.

The Court of Appeals noted that "(i)f the Commonwealth's proof was not overwhelming, neither was it thin" (*Commonwealth v. Caruso* 2014). The Court emphasised that this was not a case involving stranger identification, but rather the victim was able to recognise the defendant from his facial features and distinctive gait after having observed him for "literally hundreds of hours and having had reason to have his features etched in her memory." The order denying the motion for a new trial was affirmed (*Commonwealth v. Caruso* 2014).

2000 TO 2019

Since 2000, and particularly since 2010, there has been a significant change in all aspects of forensic gait analysis. What was once an almost accidental source of evidence has become a routine forensic option for some law enforcement agencies. Training programs, competency testing, and tested methods of analysis have all been developed. Most important was the publication of the *Code of practice for forensic gait analysis* (2019) document in the United Kingdom by the Forensic Science Regulator, which places forensic gait analysis on par with all other areas of forensic practice in terms of quality assurance and professional practice.

The first instance of an expert comparing a perpetrator's gait captured on video to that of a suspect on video and testifying in court was in 2000 in London. In the case of *R. v. Saunders* at the Central Criminal Court, consultant podiatrist Haydn Kelly identified jewellery theft suspect John Saunders as the person attempting to rob a shop, using police surveillance footage. Even though Saunders was wearing two pairs of pants, a mask, and gloves, Kelly confirmed that less than 5% of the British population had a gait similar to that of Saunders. Saunders was convicted (Buncombe 2000; Argus 2000).

A 2002 robbery-murder investigation in Denmark also used video to analyse a suspect's gait. The bank robbery was recorded by multiple video cameras in the bank. Niels Lynnerup and Jens Vedel were called upon to provide expert analysis of the video. They noticed immediately that one robber had a somewhat wide gait, "characterised especially by hyperextension of the knee joints, as well as rather outward pointing feet" (Lynnerup and Vedel 2005). They also observed a "rather pronounced swagger," with the shoulders making pronounced side-to-side movements. Lynnerup and Vedel identified similar features in the videos of the suspect. Lynnerup and Vedel opined that the suspect "might very well be identical to the perpetrator," however, they emphasised they had no basis for a statistical assessment of the degree of concordance, and that identification by their methods did not constitute identification in terms of DNA typing or fingerprinting. At trial, both prosecution and defence challenged the findings as to the degree of error and possible statistical calculations of likelihood. Nonetheless, the trial judge admitted the gait evidence and found it significant. The suspect was convicted (Lynnerup and Vedel 2005).

In *Syms v. Warden*, a prisoner appealed his conviction in Connecticut, arguing that because of his prior medical injuries, he did not have the ability to walk or run downstairs and cross a street, and as such, could not have participated in the crimes of kidnapping, robbery, conspiracy to commit robbery, and burglary (*Syms v. Warden* 2003). City officials testified that, from the scene of the crimes, there were between four and six steps to get up onto the porch, and one or two steps to go into the apartment. A doctor was asked whether the defendant could have, in escaping, run down four stairs. He stated that the defendant could have negotiated the stairs and could have run from the scene to Main Street, a distance of 354 feet, but would have done it with an altered gait. The state countered with testimony from the Medical Director for the Department of Corrections who said that Syms was "quite capable of performing the activities of going down the stairs and running away" (*Syms v. Warden* 2003). Syms' appeal was denied.

In the United Kingdom, 36-year-old John Gibson Rigg was caught on CCTV cameras carrying a laptop computer and plasma TV stolen from a house in Lancashire. Police found Rigg's DNA on a screwdriver left outside the property during the burglary in January 2007 and arrested him. While in custody, police officers noticed he had a distinctive walk, much like that of a man pictured in CCTV recovered from near the crime scene. Police enlisted the assistance of podiatrist Ian Linane, who specialised in gait analysis. He confirmed there were significant similarities (*Lancashire Telegraph* 2008). Rigg pled guilty and was sentenced to two years in jail. After the case, Linane commented:

> The analysis consists of close inspection of the movement, posture and gait of the unknown individual with footage of a known suspect, and a comparison between the two being made. Such comparisons of gait,

posture and movement have been used since the year 2000 in both low and high-profile cases... This type of approach can be valuable in a number of ways in that it is not always hindered by types of clothing worn, or even disguises.

(Lancashire Telegraph **2008)**

Detective Constable Mark Cruise also remarked, "I hope this sends out a strong message that even criminals who conceal their faces can still be identified due to this relatively new type of forensic analysis" (*Lancashire Telegraph* 2008).

British podiatrist Haydn Kelly again provided expert analysis of gait evidence at a murder trial in Victoria, British Columbia. Kelly told the Canadian jury that there were similarities in the gait exhibited by the perpetrator captured on video and the gait of Daniel Aitken, the accused, seen on video footage. Kelly told the jury that there was a "very strong likeness between the very abducted feet of the shooter in the video and the person made known to me by the RCMP as Daniel Aitken." Kelly explained that abduction is defined as the way in which a foot moves away from the middle of the body. Kelly also opined that there was a strong likeness of the "everted" left foot on the person in the video and Aitken (*R. v. Aitken* 2008). Aitken was convicted, and on appeal he argued that the judge should have ruled the gait expert's testimony inadmissible. He challenged the admissibility of the gait evidence on the grounds that it lacked the requisite level of reliability for novel science, which must be subjected to special scrutiny. It was suggested that the forensic gait analysis provided by Kelly could be properly considered "novel" because it was "the first time that such evidence has been advanced in a Canadian court" (Dickson 2008; *R. v. Aitken* 2012). Justice Hall noted that the relationship between podiatry, clinical gait analysis, and forensic gait analysis was described by the trial judge:

+ Podiatry is the study, diagnosis, and management of conditions affecting the foot. The field of study is an ancient one, stretching back a thousand years. Gait analysis is the analysis of the style or manner in which a person walks, sometimes because of symptoms or troubling pathology (*R. v. Aitken* 2012).
+ Forensic gait analysis is the term used to describe the application of gait analysis knowledge to legal problems (*R. v. Aitken* 2012).

The British Columbia Court of Appeal observed that Aitken was, in effect, arguing that the evidence failed to satisfy the factors outlined in the U.S. Supreme Court decision of *Daubert v. Merrell Dow Pharmaceuticals Inc.* (1993; *R. v. Aitken* 2012), which the Supreme Court of Canada acknowledged in a 2000 decision could be "helpful in evaluating the soundness of novel science" (*R. v. J.-L.J.* 2000; *R. v. Aitken* 2012). Aitken's attorneys were looking to have the court impose a more rigorous standard on the expert testimony, thereby hoping to have it declared inadmissible. However, the Court noted that in evaluating Kelly's qualifications as an expert witness, the trial judge rejected the argument that his evidence was novel science. The trial judge observed, "Podiatry has been in existence for a thousand years and the expertise of a podiatrist to analyse an individual's gait has long been accepted and practised in a clinical setting" (*R. v. Aitken* 2012). Justice Hall wrote that implicit in this conclusion is a determination that forensic applications of podiatry and gait analysis do not render the practice "novel" for the purposes of the *Mohan* test (*R. v. Mohan* 1994; *R. v. Aitken* 2012). That case identified the following criteria for the admissibility of expert opinion evidence:

+ relevance
+ necessity in assisting the trier of fact
+ the absence of any exclusionary rule
+ a properly qualified expert (*R. v. Mohan* 1994)

Justice Hall found that the trial judge did not err in so holding, citing the *Otway* decision (described below, a case decided in the interim before Aitken's appeal) as persuasive authority supporting the admissibility of forensic gait analysis in this case (*R. v. Aitken* 2012).

In *Otway*, a suspect was convicted in the United Kingdom in 2009 using gait analysis from a video recording. CCTV footage showed a man at a service station walking from a vehicle that was later used in a murder (*R. v. Otway* 2011). Elroy Otway was seen in the video and was identified, not by his face, but by his gait. Identification was made by David Blake, a podiatrist, who examined the CCTV footage and compared it to video of Otway taken at a police station. Otway's defence counsel contested the admissibility of the testimony of Blake for the following reasons:

+ the image comparison was within the purview of the jury themselves, and expert testimony wasn't required
+ a lack of any statistical database on which to evaluate the significance of any opinion
+ no scientific basis or measurements to support the expert's methodology
+ a lack of a "sufficiently recognised body" of experience to support gait analysis (*R. v. Otway* 2011; Adam 2016)

However, Blake's expert report provided justification and validation of the scientific value of the field. This convinced the judge that gait analysis was a sufficiently organised body of knowledge, that Blake was qualified to provide expert opinion, and that the expert was knowledgeable as to the range of both common and more unusual gait characteristics (*R. v. Otway* 2011).

The judge rejected Otway's challenge and permitted Blake to opine on the similarities in gait between Otway and those of the man in the CCTV images to assist the jury in their evaluation of the evidence. Otway was convicted of murder.

On appeal, Otway made the same arguments, and the appellate court affirmed the verdict. He also contended that, while Blake was qualified to identify the characteristics of an individual's gait, he was not qualified to compare them between two people. The Court dismissed this notion as untenable. However, the parameters of expert opinion on gait analysis were established. The Court noted:

> We agree…however, that Mr. Blake's ability safely to express his ultimate conclusion in terms of probability or a match, even probability based on Mr. Blake's clinical experience, was insufficiently established. It is important that juries are not misled to an over-valuation of comparison evidence.
>
> **(*R. v. Otway* 2011)**

The appellate court stressed that, generally speaking, the admissibility of such "podiatric" evidence should be handled on an individual basis, to be certain that the testimony is limited to the area of expertise of the particular expert and that the court clearly sets the framework for the scope of the evaluation (*R. v. Otway* 2011).

The Connecticut Court of Appeals reviewed the claims raised by a defendant who alleged that the trial court improperly admitted certain evidence in his DUI conviction (*State v. Coyne* 2010). While in this 2010 decision the appellate court found that it did not need to discuss the merits of the evidentiary claims because the defendant failed to provide any meaningful analysis of harm, it is notable that again, gait analysis was part of the defence strategy. At trial, a non-practising podiatrist working for a "forensic gait analysis group" testified that the defendant had come to him for a gait analysis in connection with the drunk driving charge. The podiatrist's assessment entailed visual analysis, field sobriety testing, and a computerised gait analysis. The expert testified that on the basis of the results of this analysis, he concluded that the defendant had a problem with his gait that would prevent him from performing the field sobriety tests. However, the podiatrist could not connect the defendant's gait issues with the inculpatory results of the horizontal gaze nystagmus test, his inability to drive or to count properly, his bloodshot eyes, his confusion, or the odour of alcohol on his breath. After the conclusion of evidence, the jury returned a guilty verdict. Based on this evidence, the jury's verdict was affirmed (*State v. Coyne* 2010). It was an original approach by the defence, but one that was ultimately unsuccessful.

In a 2012 Oregon case, a trial court denied the defendant's request to demonstrate to the jury his supposedly peculiar gait so that they could compare it to the gait of the man shown on a surveillance video taking a gun from the victim's vehicle (*State v. Fivecoats* 2012). The Court of Appeals held that it was an error for the trial court to rule that the defendant demonstrating his walk would have been *testimonial*, so as to waive his right against self-incrimination because walking was not testimonial. At trial, the transcript read as follows:

> THE COURT: Well, first of all, are you trying to prove a gait? Are you trying to show that he has some type of physical—
> (DEFENDANT'S COUNSEL): Yes, Your Honor.
> THE COURT: —something that prevents him from walking? Because that would affect how far he needs to walk. Instead of telling the Court you want your client to walk, you would need to give more information.
> (DEFENDANT'S COUNSEL): Your Honor, I believe that if he walked from, say, here to the bench and back, that would be sufficient to show he's had a broken back. He has—his gait is not a regular gait. And I believe that it is important for them to see how he walks as compared to the person that's on the video.
>
> **(*State v. Fivecoats* 2012)**

The trial judge at first agreed that the defendant could walk from the counsel table to the bench and back. However, when it came to the court's attention that the defendant was invoking his right against self-incrimination by not testifying, the judge ruled that, if the defendant did so, he could not walk because the walk would be testimony. The court stated that the defendant could not "get[ ] up and…walk[ ] in front of the jury if he's not going to take the stand. Now, if he's going to take the stand, then he can walk. But you can't have it both ways" (*State v. Fivecoats* 2012). The prosecution played the surveillance video for the jury, and two officers testified about their independent identifications of the defendant as the man in the video. One officer testified that she had known the defendant, whom she described as a distinctive looking person, for five or six years and that she recognised his mannerisms and "twitchy walk," which she described as "not a normal gait" (*State v. Fivecoats* 2012). The second officer testified that he, too, had known the defendant for five years, and he described the defendant as having a distinctive face and jerky movements. The defendant's only witness testified that she had known the defendant his entire life and that the defendant had suffered a neck injury that affected his ability to move, making him walk with a pronounced gait, "like one leg's shorter than the other" (*State v. Fivecoats* 2012). When asked by the prosecution if the defendant had "jerky movements" in addition to a limp, she agreed. The jury ultimately convicted the defendant on all counts. However, the Oregon Court of Appeals agreed with the defendant's argument about demonstrating his walk, and while the evidence against him was substantial, it was conceivable that his walk was "so distinctive and so distinctively different from the walk on the brief video that a juror could be persuaded that the defendant was not that man" (*State v. Fivecoats* 2012).

In 2015, the first observational gait analysis tool designed specifically to be used in the forensic context, the Sheffield Features of Gait Tool (see Appendix 1), was introduced and accepted as evidence in court in the United Kingdom. The Tool provides a systematic methodology to assist with gait analysis that has been tested for both repeatability and reproducibility (Birch et al. 2019).

A 2016 U.S. case involving the perpetrator's gait, *Michigan v. Ballard* (2016), arose from an armed robbery that occurred at a Detroit gas station. After the gas station attendant and manager identified a possible suspect as the masked robber, the suspect was arrested and charged with armed robbery. At trial, the victim of the robbery and the gas station attendant identified the suspect as the perpetrator. In addition, a surveillance video that captured the robbery was admitted into evidence at trial. The video showed that the robber had a limp. Based on the robber's limp as shown in the video, the gas station manager, who had not witnessed the robbery, also identified the defendant as the robber seen in the video (*Michigan v. Ballard* 2016). At trial, the defendant presented a mistaken identity defence. Along with an alibi, the defendant acknowledged that he had a limp due to a right leg amputation; however, his limp was different than the perpetrator's limp. Nonetheless, the trial court found the eyewitness identifications credible and found him guilty (*Michigan v. Ballard* 2016). On appeal, the Michigan Court of Appeals found that the two eyewitnesses to the crime and their credibility were crucial to the outcome of the case. The Court said that Ballard's defence counsel's performance fell below an objective standard of reasonableness when the attorney failed to utilise readily available video gait evidence to impeach their identifications of the defendant at trial (*Michigan v. Ballard* 2016). The defendant's attorney conceded that the "(defendant)'s gait related or compared to the gait of the perpetrator as seen in the surveillance video was a pivotal issue in the trial," but admitted that he did not consider consulting with an expert in gait analysis and that he did not do "any searching for a gait expert or anything of that nature" (*Michigan v. Ballard* 2016). The Court of Appeals reversed the trial court's denial of the defendant's motion for a new trial and vacated the defendant's conviction. The defendant's counsel was at fault for failing to use the video of the gait and analysis by an expert to impeach the eyewitnesses. This deprived the defendant of a substantial defence, the appellate court reasoned (*Michigan v. Ballard* 2016). The comparison between the perpetrator's gait and defendant's gait was central to this case, the Court of Appeals noted. The defendant's counsel's "complete failure to investigate, consult, or retain an expert on…gait analysis was objectively unreasonable" (*Michigan v. Ballard* 2016). As a result, the defendant was denied the effective assistance of counsel, and the trial court therefore abused its discretion in denying his motion for a new trial (*Michigan v. Ballard* 2016).

Both gait expert opinion and eyewitness lay testimony have become more common in American courts, particularly in criminal cases. Testimony of a defendant's gait has been accepted by many judges as viable and admissible evidence. To that end, several recent criminal convictions in the United States were upheld on appeal with gait evidence in the form of video footage and eyewitness identification.

In 2018, a convicted thief claimed that the outcome of the trial would have been different had a witness not testified that he was able to identify the defendant by "gait" on the surveillance video. The court affirmed his conviction (*People v. Beasley* 2018).

In a 2019 Texas case, the evidence was sufficient to support the defendant's conviction of arson where witnesses testified that they saw him fleeing from the arson when it occurred, and one witness testified that she recognised him from the surveillance recordings showing him pulling out of his parking space and driving away (*Gibson v. Texas* 2019).

The jury heard a witness testify that from watching the surveillance recordings, she knew that the defendant had set the fire. She stated that from the recordings, she identified him through his gait. She testified, "I've known him for years, and I know how he walks." She stated, "His belly pops out like that, and he just stumbles like that, like a little duck walk." She testified that she had "[n]o doubt at all" that the man in the video was Gibson (*Gibson v. Texas* 2019). The Court of Appeals held that the jury could have "rationally accepted" the witness' identification of the defendant through the recordings. Although the defendant emphasised that the recordings did not show the arsonist's face, the witness did not identify him by his face. Instead, she identified him by his gait, which the recordings depicted (*Gibson v. Texas* 2019).

Nathan Kinney was convicted of armed assault with intent to murder, two counts of aggravated assault and battery by means of a dangerous weapon, and unlawful possession of a firearm, in connection with the 2008 shooting of two men outside a Boston bar (*Commonwealth v. Kinney* 2019). The defence challenged the contention that the defendant was the shooter. The evidence at trial consisted primarily of video surveillance recordings from both inside and outside the bar that evening, and the identification of the defendant by multiple witnesses. A police officer testified that he had been called to the bar later that evening, after the shooting, and was shown surveillance videos of the man that employees believed was involved in the shooting. That evening, and at trial, the officer identified the defendant on video entering the bar prior to the shooting, after recognising his face, the jacket, and the neck tattoo, which was visible on the video. At trial, the officer also viewed a different video, the video from outside the bar showing the events leading up to and including the shooting, and testified that the defendant was the person seen exiting the car and walking towards the bar entrance. Defence counsel objected and the judge struck part of the officer's testimony, disallowing the officer from stating the person on the video was the defendant. However, the officer was permitted to testify that the person shown on the video outside the bar had a similar stature and gait to that of the defendant, and was wearing the same jacket that the officer had seen on the defendant that evening. The Massachusetts Court of Appeals held that this testimony was admissible as a simple lay opinion, based on the officer's observations of the defendant in numerous prior interactions, including earlier that evening (*Commonwealth v. Kinney* 2019).

In another 2019 case, a California jury found defendant Jeremy Charles Pratt guilty of felony vandalism (*People v. Pratt* 2019). He appealed, arguing that there was insufficient evidence of his identity as the vandal to support his conviction. In this case, the evidence, including the video footage viewed by the jury and testimony of witnesses who knew and recognised the defendant, showed him leaving his workplace and driving away that morning in his pick-up truck, walking onto the school campus and into the parking lot, walking away from the school and turning the corner, and then later arriving back at his workplace. The witnesses described Pratt's build, distinctive gait, and uncommon hair colour, as well as the jacket he often wore. The Court of Appeals found that the jury could evaluate whether he did, in fact, have a distinctive gait that would allow those who knew him to recognise him from the rear (*People v. Pratt* 2019).

GAIT EVIDENCE IN CIVIL PROCEEDINGS

Gait evidence and testimony by experts has also been admitted in insurance, worker's compensation, and other civil matters. Medical expert testimony can be put forth to prove or disprove an inconsistency with medical issues that have been claimed by an individual seeking benefits or compensation from his or her employer or insurer.

An Ohio man took intermittent leave for a chronic back condition and was fired for abusing his leave. He brought a retaliation claim under the Family and Medical Leave Act (FMLA) (*Tillman v. Ohio Bell Tel. Co.* 2011). That claim failed because the employer held an honest belief that the employee Tillman had abused his FMLA leave and violated the company policies. He frequently exercised his FMLA leave time on weekends or adjacent to scheduled days off, and he "forecasted" his leave days in advance. This caused the employer to become suspicious. The employer had surveillance video that was analysed by a medical expert. The expert found, in part, that Tillman "did not limp or favor a leg when walking," and "[h]is gait was smooth, easy, and symmetric with normal arm swing" (*Tillman v. Ohio Bell Tel. Co.* 2011). The District Court dismissed Tillman's claim, and based on this evidence, the Sixth Circuit Court of Appeals affirmed the decision.

Gait analysis in alleged medical fraud matters may be important in determining if a claimant were capable of walking based on the examination of the wear pattern on his shoes. This analysis was detailed by Canadian podiatrist Norman Gunn, who was called upon to review the case of a claimant who maintained he was paralysed and wheelchair-bound due to an accident. Gunn was charged with determining if the plaintiff could walk. The claimant explained that the outsole wear patterns on his shoes were caused by resting his feet on his wheelchair platform and made worse by his wife sitting on his knee. Gunn found that the wheelchair platform was not long enough to cause the anterior wear pattern on the claimant's outsole. The claimant responded that this wear was due to pressing against doors with his feet extended while being pushed in the wheelchair. Gunn tested that contention by comparing the wear patterns on the man's shoes with those seen on the shoes of other wheelchair patients. His analysis and comparison bore out that the claimant was not wheelchair-bound as he asserted (Nirenberg, Vernon, and Birch 2018).

DISCUSSION

The cases described in this chapter show that gait has been established as admissible evidence. The genesis of the application is found in criminal proceedings in which lay people gave eyewitness testimony opining that they recognised an individual by their gait. This has developed into an area of specialisation, with expert witness testimony in which technical and scientific knowledge has been leveraged to assist both prosecution and defence. In general terms, the historical use of gait analysis as evidence has been of four types:

- non-expert, eyewitness testimony
- expert opinion
- expert specialised knowledge/technical opinion
- expert scientific opinion

Although expert reporting and testimony on gait analysis may be expected to be more reliable, some courts continue to admit non-expert or lay witness gait evidence, a lay person who provides non-expert gait analysis being considered to be an "ad hoc expert" (Edmond et al. 2013; Edmond, Martire, and San Roque 2017). Prosecutors continue to play video footage showing gait to juries and eyewitnesses who have identified defendants by the way they walk at trial (*Commonwealth v. Caruso* 2014; *State v. Richardson* 2017; *People v. Beasley* 2018; *Commonwealth v. Kinney* 2019; *Gibson v. Texas* 2019; *People v. Pratt* 2019), despite evidence that doing so may have a more complex effect on jury decision making that may previously have been thought (Montepare, Goldstein, and Clausen 1987; Birch, Birch, and Bray 2016; Caruso, Burns, and Converse 2016).

The cases described in this chapter demonstrate that initial involvement of experts did not coincide with the use of currently accepted scientific methodologies. Rather the evidence put forth appearing to have been based primarily on the specialised knowledge of the expert involved. The *Jerry Mark* case (*State v. Mark* 1980) is a distinct exception. In that murder trial, the expert, Lichty, arguably employed an ACE-V (Analysis, Comparison, Evaluation, and Verification) approach, a strategy that is now commonly applied in the forensic sciences (Reznicek, Ruth, and Schilens 2010). While consideration of the cases did not reveal the specifics of the methodologies used by the experts in each case, the adoption of an ACE-V approach by practitioners undertaking work involving gait analysis as evidence appears to have been a more recent development.

The *Saunders* case ushered in a new phase in the use of forensic gait analysis, involving the analysis and comparison of gait from CCTV recordings (Buncombe 2000). Although this process had been undertaken earlier in 1996 (*State v. Williams* 1996), *Saunders* was the first time an expert witness had been engaged to perform such analysis, and the first case in which statistical data had been presented as part of the gait expert's opinion.

The cases described also demonstrate the use of a defendant's appeal of ineffective assistance of counsel when their attorney failed to employ gait analysis evidence to impeach witness identifications based on gait. To establish ineffective assistance of counsel, a defendant must prove his attorney's representation fell below an objective standard of reasonableness; and but for the attorney's errors, there is a reasonable probability that the result of the trial would have been different (*Strickland v. Washington* 1984). In *Ballard* (*Michigan v. Ballard* 2016), the defendant's trial counsel did not impeach the testimony of two eyewitnesses to the robbery when gait evidence was readily available. The defendant appointed a new appelate counsel who subsequently offered the testimony of an expert in the field of orthotics and prosthetics who was able to conclude that the perpetrator's gait was inconsistent with the defendant's. The gait expert opined that the video showed that the perpetrator had an injury or weakness in his *left* leg, rather than (like defendant) in his *right* leg, and that it would be very difficult, if not impossible, for the defendant to move in the manner in which the perpetrator moved in the video. Again, the defendant's trial counsel admitted that he failed to consider consulting an expert in gait analysis. Whether this is a trend in defence strategy is unclear; however, in the one case where the defendant was successful on appeal, an expert in gait analysis was available, and the attorney failed to use that testimony at trial (*Michigan v. Ballard* 2016).

Gait has been used as evidence for over one hundred years, and its use continues to become more widely accepted in courtrooms. As the body of research informing and underpinning professional practice grows, so too does the use of more unified and standardised approaches to gait analysis and comparison (Birch et al. 2013; Birch et al. 2014; Forensic Science Regulator 2017; Birch et al. 2019). The development of professional standards and the alignment of forensic gait analysis practice with other forensic disciplines are further enhancing the use of gait as evidence (*Code of practice for forensic gait analysis* 2019[1]).

NOTE

1. Currently available from the Office of the Forensic Science Regulator, UK.

REFERENCES

Adam, C. *Forensic Evidence in Court: Evaluation and Scientific Opinion*. Hoboken, NJ: John Wiley & Sons, 2016.

Appel, T. "A Woman Who Claims Night Stalker Suspect Richard Ramirez…" *United Press International Inc.*, September 10, 1987.

"Bandy Bandit Is Caught on Security Cameras." *The Argus*, July 12, 2000, Sussex, England.

Bertillon, Alphonse. *Signaletic Instructions Including the Theory and Practice of Anthropometrical Identification*. Chicago: Werner Company, 1896.

Biographies, Alphonse Bertillon (1853–1914), *Visible Proofs: Forensic Views of the Body*. Bethesda, MD: U.S. National Library of Medicine, December 2014. Accessed July 2019. www.nlm.nih.gov/visibleproofs/galleries/biographies/bertillon.html.

Birch, I., L. Raymond, A. Christou, M. A. Fernando, N. Harrison, and F. Paul. "The Identification of Individuals by Observational Gait Analysis Using Closed Circuit Television Footage." *Science and Justice* 53(3) (2013):339–342.

Birch, I., W. Vernon, G. Burrow, and J. Walker. "The Effect of Frame Rate on the Ability of Experienced Gait Analysts to Identify Characteristics of Gait from Closed Circuit Television Footage." *Science and Justice* 54(2) (2014):159–163.

Birch, I., T. Birch, and D. Bray. "The Identification of Emotions from Gait." *Science and Justice* 56(5) (2016):351–356.

Birch, I. "The Development and Testing of the Sheffield Features of Gait." *International Association for Identification Annual International Educational Conference*, San Antonio, TX, 2018.

Birch, Ivan, Maria Birch, Lucy Rutler, Sarah Brown, Libertad Rodriguez Burgos, Bert Otten, and Mickey Wiedemeijer. "The Repeatability and Reproducibility of the Sheffield Features of Gait Tool." *Science and Justice* 59(5) (2019):544–551.

Buncombe, A. "Gang Leader Is Unmasked by His Bandy-Legged Gait." *The Independent*, July 13, 2000, London.

Caruso, E. M., Z. C. Burns, and B. A. Converse. "Slow Motion Increases Perceived Intent." *Proceedings of the National Academy of Sciences* 113(33) (2016):9250–9255.

Chartered Society of Forensic Sciences and College of Podiatry in association with the Forensic Science Regulator. 2019. *Code of practice for forensic gait analysis*, Issue 1. Birmingham: The Forensic Science Regulator.

Commonwealth v. Caruso, 4 N.E.3d 1283 (Mass. App. 2014).

References

Commonwealth v. Kinney, 2019 Mass. App. Unpub. LEXIS 198 (Ma. App. 2019).
Commonwealth v. Spencer, 39 A.2d 820 (Pa. Supr. 1994).
Daubert v. Merrell Dow Pharmaceuticals, Inc., 509 U.S. 579 (1993).
Dickson, L. "Gait Analysis Explained at Victoria Murder Trial." *Times Colonist*, November 19, 2008.
Doyle, A. C. *The Case of Oscar Slater*. New York: Hodder & Stoughton, George H. Doran Company, 1912.
Edmond, G., S. Cole, E. Cunliffe, and A. Roberts. "Admissibility Compared: The Reception of Incriminating Expert Evidence (i.e., Forensic Science) in Four Adversarial Jurisdictions." *University of Denver Criminal Law Review* 3 (2013):31–109.
Edmond, G., K. Martire, and M. San Roque. "Expert Reports and the Forensic Sciences." *University of New South Wales Law Journal* 40 (2017):590.
Federal Rule of Evidence 701.
Forensic Science Regulator. *Codes of Practice and Conduct: For Forensic Science Providers and Practitioners in the Criminal Justice System, Issue 4*. Birmingham: The Forensic Science Regulator, 2017.
Gibson v. Texas, 2019 Tex. App. LEXIS 1558 (Tex. App. 2019).
Lynnerup, N., and J. Vedel. "Person Identification by Gait Analysis and Photogrammetry." *Journal of Forensic Sciences* 50(1) (2005):112–118.
Michigan v. Ballard, 2016 Mich. App. LEXIS 1550 (Mi. App. 2016).
Montepare, J. M., S. B. Goldstein, and A. Clausen. "The Identification of Emotions from Gait Information." *Journal of Nonverbal Behavior* 11(1) (1987):33–42.
Nirenberg, M., W. Vernon, and I. Birch. "A Review of the Historical Use and Criticisms of Gait Analysis Evidence." *Science and Justice* 58(4) (2018):292–298.
Old Bailey Proceedings Online. "Thomas Jackson." *The Proceedings of the Old Bailey*, December 16, 1839. Accessed 31/01/19. https://www.old baileyonline.org/browse.jsp?div=t18 391216-382.
People v. Beasley, 2018 Ill. App. Unpub. LEXIS 1238 (Il. App. 2018).
People v. Colbert, 2010 Cal. App. LEXIS 1348 (Cal. App. 2010).
People v. Pratt, 2019 Cal. App. Unpub. LEXIS 2051 (Cal App. 2019).
R. v. Aitken 2008, BCSC 1423.
R. v. Aitken 2012 BCCA 134.
R. v. J.-L.J. 2000 SCC 51.
R. v. Mohan 1994 CanLII 80.
R. v. Otway 2011, EWCA Crim 3.
Reznicek, M., R. M. Ruth., and D. M. Schilens. "ACE-V and the Scientific Method." *Journal of Forensic Identification* 60 (2010):87–103.
Smith, Sir S. *Mostly Murder*. Toronto: Clark, Irwin and Co., 1959.
State v. Coyne, 985 A.2d 1091 (Ct. App. 2010).
State v. Fivecoats, 284 P.3d 1225 (Or. App. 2012).
State v. Hills, 241 La. 345 (La. 1961).
State v. Iverson, 187 N.W.2d 1 (N.D. 1971).

State v. Mark, 286 N.W.2d 396 (Iowa 1980).
State v. Richardson, 2017 Wash. App. LEXIS 2524 (Wa. App. 2017).
State v. Santos, 935 A.2d 212 (Ct. App. 2007).
State v. Williams, 558 N.W.2d 705 (Wi. App. 1996).
Strickland v. Washington, 466 U.S. 668 (1984).
Syms v. Warden, 2003 Conn. Super. LEXIS 73 (Ct. Supr. 2003).
Tillman v. Ohio Bell Tel. Co., 2011 U.S. Dist. LEXIS 74329 (N.D. Ohio 2011).
United States ex rel. Bass v. Ahitow, 1998 U.S. Dist. LEXIS 17096 (N.D. Ill. 1998).
"Video: 'Swaggering' Burglar Caught Out by His Own Walk." *The Lancashire Telegraph*, April 11, 2008, Lancaster, England.

Fundamentals of human gait and gait analysis

Ambreen Chohan, Jim Richards and David Levine

Locomotion is the process of moving from one location to another, and necessitates starting, stopping, controlling and modifying speed and direction, and adapting to one's environment. Whilst the locomotive process follows a basic rhythmic pattern, the detail of "normal" or "usual" locomotion needs to be understood before one can decipher that which is abnormal or unusual.

Human locomotion is usually based on walking, which is a process involving two legs alternately providing support and propulsion of the body in space, with at least one lower limb remaining in contact with the ground at all times. This makes it distinct from running, in which there is a period when neither foot is in contact with the ground. Gait is the manner or style in which a person undertakes a locomotor activity such as walking or running.

Gait analysis has evolved considerably over the last 100 years from descriptive studies in the early 1900s to recent multifaceted techniques involving complex mathematical modelling and analysis (Garrison 1929, Bresler and Frankel 1950, Steindler 1953, Murray 1967, Inman et al. 1981, Sutherland 2001, 2002, 2005).

A thorough understanding of the human gait cycle is a fundamental prerequisite for anyone intending to undertake forensic gait analysis, and useful knowledge for those involved in its commissioning and application.

THE GAIT CYCLE

The gait cycle is defined as the time between two successive occurrences of one of the repetitive events involved in walking (Levine, Richards, and Whittle 2012). It is usually determined as the period from initial contact of the foot with the ground to the next initial contact of the same foot. The gait cycle can be divided into two major phases, the *stance phase* when the foot is in contact with the ground, and the *swing phase* when the foot is in the air.

The timing of single and double limb support during a gait cycle, between initial contact of one limb and toe off of the opposite limb, is shown in Figure 3.1. During a "normal" gait cycle, initial contact with the ground by the right foot occurs whilst the left lower limb is still in contact with the ground, resulting in the *double support phase* or *double support stance*, the period when both feet are in contact with the ground. When left toe off occurs, the left lower limb enters the swing phase, and the right lower limb remains in contact with the ground, resulting in the right *single limb support phase* or *single support stance*, which occurs until the left foot comes

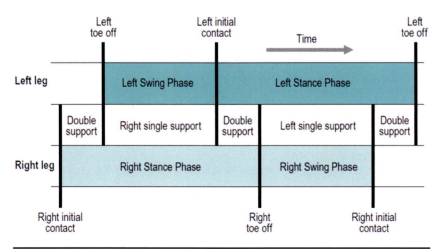

FIGURE 3.1 Timing of single and double support in the gait cycle (Levine, Richards, and Whittle 2012).

into contact with the ground once more. There is then another period of double limb support, which occurs until the right foot leaves the ground at toe off, ending the right stance phase. The left single limb support phase then occurs during the right swing phase. This phase ends when the right foot recontacts the ground, initiating another phase of double limb support.

At initial contact, one foot is placed ahead of the other, and this lower limb is then described as the *leading leg*, the other being referred to as the *trailing leg*. The position of lower limbs during gait is described in more detail using the gait cycle wheel in Figure 3.2.

In this diagram (Figure 3.2), the lower limb position during a single gait cycle is broken down into seven key events, four of which occur during stance phase (support phase) and three occurring in swing phase. The major events of the gait cycle are:

1. Initial contact
2. Opposite toe off
3. Heel rise

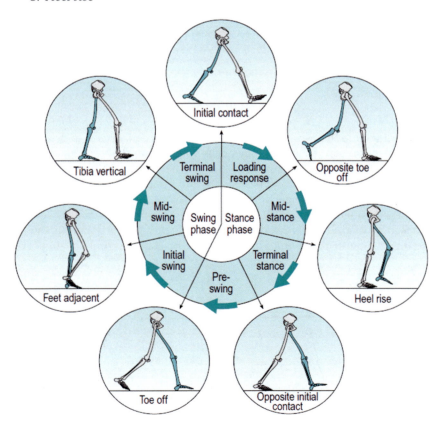

FIGURE 3.2 The gait cycle (Levine, Richards, and Whittle 2012).

4. Opposite initial contact
5. Toe off
6. Feet adjacent
7. Tibia vertical
(See Figure 3.3.)

These events are used to define the phases of the gait cycle seen in Figure 3.2. During the stance phase, between initial contact of the lower limb and toe off of the same lower limb, four phases occur:

1. Loading response
2. Mid-stance
3. Terminal stance
4. Pre-swing

The time taken to complete the stance phase of the gait cycle is known as *stance time*. The stance phase lasts approximately 60% of the gait cycle. The swing phase lasts approximately 40% of the gait cycle, occurring from toe off of the lower limb to initial contact of the same lower limb. During swing, three phases occur:

1. Initial swing
2. Mid-swing
3. Terminal swing

The time taken to a complete gait cycle is known as the *cycle time*. It is important to note that while these events and phases can be used to describe normal or usual gait, shortfalls may occur when using this terminology to describe pathological or unusual gait.

During a complete gait cycle there are two periods of double limb support and two periods of single limb support. Each period of double limb support lasts approximately 10% of the gait cycle, although this can change if the speed of walking is increased, as the limbs spend a proportionally greater amount of time in the swing phase than in the stance or double support

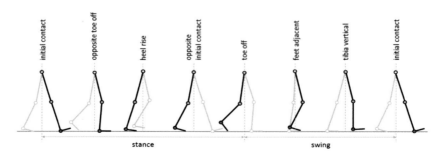

FIGURE 3.3 Events of the gait cycle.

phases (Murray 1967). With the increase in gait speed, the double support phase shortens and eventually disappears, defining the point at which a person makes the transition from walking to running. During running, non-supported phases occur when both limbs are in a *flight phase* and not in contact with the ground.

THE GAIT CYCLE IN DETAIL

EVENT 1: INITIAL CONTACT

An event frequently known as heel strike or initial foot contact in normal gait, initial contact is the beginning of the loading response phase.

- **Ankle:** At initial contact, the ankle is close to a neutral position (neither plantarflexed nor dorsiflexed) with the foot pointing upwards as a result of the tibia sloping backwards. The foot is also usually slightly inverted. At initial contact, the dorsiflexor muscles in the anterior of the leg act eccentrically to control plantarflexion of the foot.
- **Knee:** During late swing, and before initial contact, the knee extends and is fully, or close to fully, extended before initial contact. This extension of the knee is controlled by a combination of the quadriceps and the hamstring muscles which prevents hyperextension of the knee. Following initial contact, the knee flexes under maximum weight-bearing, helping to absorb the shock of contact with the ground.
- **Hip:** Around the time of initial contact the gluteus maximus begins to contract, initiating the extension of the hip, which completes around the time of initial contact of the opposite lower limb.
- **Upper body:** The trunk is approximately half a stride length behind the leading leg at initial contact, and is moving laterally towards the side of the leading leg. The pelvis is rotated slightly forward to facilitate hip extension of the lower limb making contact with the ground, and the opposite shoulder and arm are extended forwards. Arm swing patterns can vary greatly both within and between individuals, but excursion generally increases with the speed of walking.

LOADING RESPONSE

The loading response phase is defined as the period between initial contact and opposite toe off, and is a period of double support, both lower limbs being in contact with the ground.

- **Ankle:** After initial contact, which creates a pivot point at the rear foot, the foot is lowered to the ground by plantarflexion of the ankle. This is known as the loading response, during which

ankle plantarflexion is controlled by eccentric contraction of the tibialis anterior muscle. The loading response can typically last for the first 10–12% of the gait cycle.
+ **Knee:** From being fully or almost fully extended at initial contact, the knee flexes during double support phase, initiating stance phase flexion with the contraction of the quadriceps controlling the magnitude and speed of flexion.
+ **Hip:** The hip continues to extend during this part of the gait cycle, through concentric contraction of the hip extensors, gluteus maximus and hamstrings.
+ **Upper body:** The trunk is at its lowest vertical position during double support and continues to move laterally towards the side of the leading leg. The arms reach their maximum anterior-posterior swing position.

Event 2: Opposite toe off

Opposite toe off is the end of the period of double support and the beginning of mid-stance, the first phase of single limb support. The foot of the leading leg reaches the foot flat position around the time of opposite toe off. Opposite toe off of the trailing leg, which occurs at around 10% of the gait cycle, marks the end of stance phase and the beginning of swing phase for this lower limb.

+ **Ankle:** As soon as the leading foot is flat on the ground, the motion of the ankle changes from plantarflexion to dorsiflexion as the tibia moves over the stationary foot. Pronation[1] of the foot and internal rotation of the tibia reach peak values around the time of opposite toe off, and then begin to reverse. These movements occur together due to the anatomy of the ankle and subtalar joints (Inman et al. 1981, Rose and Gamble 1994).
+ **Knee:** The knee continues to flex and reaches peak flexion during early mid-stance, before beginning to extend. This flexion is controlled by the quadriceps muscles, which allow the knee to act like a spring, stopping the vertical force from building too rapidly (Perry 1974). The degree to which the knee flexes is influenced by walking speed and decreases in slower walking.
+ **Hip:** At opposite toe off, the hip is continuing to extend due to contraction of the gluteal and hamstring muscles.
+ **Upper body:** Having reached their most extreme positions, the arms begin to move in the opposite directions. The pelvis on the leading side starts to rotate back towards a neutral position. The trunk starts to gain height, but loses forward speed, due to the ground reaction force.

Mid-stance

Mid-stance is the phase of the gait cycle between opposite toe off and heel rise. However, in the past, the term has been used to describe the moment when the swing lower limb passes the stance lower limb, now usually referred to as feet adjacent.

+ **Ankle:** During mid-stance and terminal stance, the tibia rotates forwards around the ankle, with the foot remaining flat on the floor, in a movement sometimes referred to as the *mid-stance rocker*. The ankle joint transitions from a plantarflexed to a dorsiflexed position, the movement being controlled by the eccentric contraction of the calf muscles, the triceps surae. The tibia externally rotates with coupled supination[2] of the foot during mid-stance and terminal stance, the supination peaking in mid-stance before pronation of the foot occurs once again.
+ **Knee:** The knee reaches its peak flexion of the stance phase at 15–20% of the gait cycle, then begins to extend again. The knee angle during mid-stance is highly dependent on the speed of walking.
+ **Hip:** The hip continues to extend transitioning from a flexed to an extended position during mid-stance. This is generally achieved by a combination of inertia and gravity. As the opposite foot leaves the ground, the pelvis is supported by the stance phase hip, the swing phase side dipping slightly.
+ **Upper body:** The trunk reaches its highest point during mid-stance. The lateral motion of the trunk also peaks during this phase as it moves towards the side of the stance lower limb. Like the feet, the arms also pass each other during mid-stance, following the motion of the opposite lower limb. The shoulders and pelvis both move through a neutral position, but in opposite directions, resulting in there being no twisting of the trunk.

Event 3: Heel rise

Heel rise or heel off marks the change from the mid-stance phase to the terminal stance phase.

+ **Ankle:** Peak dorsiflexion of the ankle is reached after heel rise. As the knee begins to flex, the ankle angle is initially maintained by the calf muscles, before the ankle moves into plantarflexion late in terminal stance. The tibia externally rotates, causing the foot to supinate. The toes remain flat on the ground as the heel rises, extension occurring at the metatarsophalangeal joints, and hindfoot inversion

is seen. The timing of heel rise varies from person to person and with walking speed.
+ **Knee:** The knee has a peak of extension close to heel rise. The mechanical effect of active plantarflexion of the ankle by the calf muscles contributes to knee extension. The proximal attachment of gastrocnemius above the knee results in the muscle not only actively plantarflexing the ankle, but also resisting hyperextension of the knee and then initiating flexion of the knee.
+ **Hip:** Hip extension is continuing at heel rise and into terminal stance, with peak hip extension occurring around the time of opposite initial contact. The hip abductor muscles contract to stabilise the pelvis, but relax before opposite initial contact.
+ **Upper body:** At heel rise, the trunk has begun its descent from its highest point reached in mid-stance. There is also a decrease in the lateral displacement of the trunk over the stance lower limb in preparation for load shifting to the opposite lower limb.

Event 4: Opposite initial contact

Opposite initial contact is the first contact of the opposite lower limb with the ground. This occurs at close to 50% of the gait cycle in a symmetrical gait, marking the end of single limb support and the onset of the pre-swing phase, the second period of double limb support. This is also known as the terminal rocker phase, the leg rotating forwards around the point of contact of the forefoot with the ground and not around the ankle joint.

+ **Ankle:** Until the foot leaves the ground at toe off, the ankle is continuing to plantarflex. The plantar fascia tightens during toe extension, and the foot moves into maximal supination, with hindfoot inversion and external tibial rotation. This results in a highly stable foot to support loadbearing.
+ **Knee:** The knee has already begun to flex by initial contact of the opposite limb, and continues to flex, with the eccentric contraction of rectus femoris and quadriceps muscles preventing it from flexing too rapidly.
+ **Hip:** The hip at opposite initial contact has reached its most extended position, and begins to flex.
+ **Upper body:** At opposite initial contact, the trunk is, as was described at initial contact, half a stride length behind the now opposite leading leg and is moving laterally towards the side of that lower limb. The pelvis is rotated slightly forward to facilitate hip extension of the lower limb making contact with the ground, and the opposite shoulder and arm are extended forwards.

Event 5: Toe off

Toe off occurs at approximately 60% of the gait cycle, and marks the transition from the stance phase to the swing phase of gait. Toe off is also sometimes referred to as terminal contact, as some pathologies can result in parts of the foot other than the toes being the last to leave the ground during walking.

+ **Ankle:** Just after toe off, the ankle reaches peak plantarflexion. The calf muscles have relaxed before toe off, and tibialis anterior now dorsiflexes the foot to clear the ground during the swing phase of gait.
+ **Knee:** By toe off, the knee has flexed to around half its peak swing phase amount. During this phase of the gait cycle the lower limb acts as a double pendulum, with knee flexion resulting from hip flexion as the foot is left behind. The eccentric contraction of rectus femoris prevents the knee from flexing excessively at higher walking speeds (Nene et al. 1999).
+ **Hip:** The hip continues to undergo flexion as the foot leaves the ground, as the result of a combination of gravity, passive soft tissue tension and muscle contraction.
+ **Trunk:** After toe off, the trunk, shoulders and arms all begin to move from their rotated positions back towards neutral positions, as the trunk rises and moves towards the side of the now supporting lower limb.

Event 6: Feet adjacent

Occurring at around 75–80% of the gait cycle, this event separates the initial and mid-swing phases. Feet adjacent occurs when the two feet are side by side, as the swinging lower limb progresses forwards, approximately halfway through the swing phase of gait.

+ **Ankle:** At feet adjacent, the ankle is progressing from its plantarflexed angle at toe off to a neutral or dorsiflexed position in terminal swing, assisting the shortening of the swinging lower limb in order to clear the ground. This shortening is largely achieved by the flexion of the knee. The foot is slightly supinated.
+ **Knee:** Knee flexion has peaked before feet adjacent and the knee is now starting to extend. As the speed of gait increases, this peak knee flexion, and time in swing phase, decrease.
+ **Hip:** Flexion of the hip continues to increase as the result of the combination of the contraction of iliopsoas and gravity.
+ **Upper body:** When the feet are adjacent, the trunk has reached its maximal vertical position over the supporting lower limb, and the arms are approximately level but moving in opposite directions.

Event 7: Tibia vertical

Tibia vertical divides the mid- and terminal swing phases of the gait cycle.

- **Ankle:** From prior to tibia vertical to the next initial contact, the ankle position is variable from slightly plantarflexed to slightly dorsiflexed. The angulation of the ankle is maintained by the continued action of tibialis anterior, which increases prior to initial contact.
- **Knee:** As tibia vertical occurs, the knee is undergoing rapid extension, moving from peak swing phase flexion to approximately full extension before the next initial contact.
- **Hip:** At approximately tibia vertical, the hip stops moving into further flexion. An increase in hamstring contraction during terminal swing limits the rate of knee extension, but keeps the hip flexed.
- **Trunk:** When the tibia is vertical the trunk has begun to descend, and is moving from peak displacement over the supporting lower limb, towards the midline.

Terminal foot contact

The end of the gait cycle is identified by the initial contact or heel strike of the same foot. To avoid confusion this is known as terminal foot contact.

Temporal and spatial parameters of gait

Parameters of gait can be divided into temporal parameters, those measured in time, and spatial, those measured in distance or angle. General gait parameters that can be used to describe the way a person walks include cycle time, cadence, speed of gait, step length and stride length (Murray et al. 1964, 1970, Rigas 1984). Figure 3.4 shows a range of suggested values for some parameters of gait.

Speed

Speed of gait can be measured by timing a person when walking a defined or known distance between two landmarks. As an example, if a person walked 100 metres in 60 seconds then the gait speed is 1.67 m/s. Observation of gait over a distance of around 10 metres can allow the determination of gait speed. During the analysis of gait, an individual is often told to walk at a natural comfortable speed over a distance several times, to remove the possible influence of conscious control and to obtain a more representative outcome. The use of the term velocity to describe speed is an incorrect use of the term unless the direction of movement is accounted for, velocity being a vector quantity.

	Age (years)	Cadence (steps/minute)	Cycle time (seconds)	Stride length (metres)	Speed (metres/second)
Female	13–14	103–150	0.80–1.17	0.99–1.55	0.90–1.62
	15–17	100–144	0.83–1.20	1.03–1.57	0.92–1.64
	18–49	98–138	0.87–1.22	1.06–1.58	0.94–1.66
	50–64	97–137	0.88–1.24	1.04–1.56	0.91–1.63
	65–80	96–136	0.88–1.25	0.94–1.46	0.80–1.52
Male	13–14	100–149	0.81–1.2	1.06–1.64	0.95–1.67
	15–17	96–142	0.85–1.25	1.15–1.75	1.03–1.75
	18–49	91–135	0.89–1.32	1.25–1.85	1.10–1.82
	50–64	82–126	0.95–1.46	1.22–1.82	0.96–1.68
	65–80	81–125	0.96–1.48	1.11–1.71	0.81–1.61

FIGURE 3.4 Suggested reference values for gait parameters in males and females.

CYCLE TIME AND CADENCE

Cycle time and cadence can be measured by counting the number of steps taken during a given period of time. During gait analysis, an individual is usually told to walk naturally over a sufficient distance that will allow a natural gait and speed to be attained. Cycle time is the duration of one complete gait cycle. Cadence is the number of steps taken per minute.

STRIDE LENGTH

Stride length can be determined directly through measurement or indirectly from speed and cycle time. One method of direct measurement is to count the strides taken when walking over a set distance. Alternatively, there are various methods of recording a series of consecutive footprints during walking, including the use of ink or talcum powder, from which a variety of measurements can be taken as shown in Figure 3.5.

Where both cycle time and speed have already been calculated, the following formulae can be used to calculate stride length in metres:

$$\text{stride length}(\text{metres}) = \text{speed}(\text{metres per second}) \times \text{cycle time}(\text{seconds})$$

OR

$$\text{stride length}(\text{metres}) = \frac{\text{speed}(\text{metres per second}) \times 2 \text{ steps} \times 60 \text{ seconds}}{\text{cadence}(\text{steps per minute})}$$

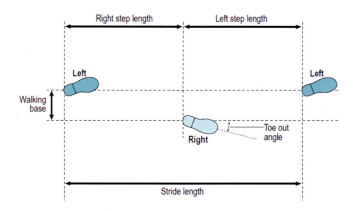

FIGURE 3.5 Foot placement during the stride (Levine, Richards, and Whittle 2012).

For accuracy, the cycle time and speed should be calculated using the same series of steps.

GAIT ANALYSIS

OBSERVATIONAL GAIT ANALYSIS WITHOUT VIDEO RECORDING

Observational gait analysis is the acquisition of kinematic information about the gait of a person from visual observations. In the last 70 years, observational analysis of gait has become an important part of clinical assessment in rehabilitation medicine, and the value of such assessment is now becoming of great interest in non-medical fields as well. The first systematic studies of gait date back to ancient times. Aristotle, Hippocrates, Galen and Leonardo da Vinci all gave useful descriptions of walking based on observation (Collado-Vázquez and Carrillo 2015). While gait analysis based on observations from video footage has obvious advantages, in some environments gait analysis based purely on direct observation is the only form of assessment available (Turnbull and Wall 1985).

Parameters of gait such as relative timing of foot contacts, the number of steps per minute, step and stride length and walking speed or velocity can be assessed from observational gait analysis, as can the relative positions of limbs and the angulation of joints. These observations can be used to assess the symmetry, or lack of symmetry, between left and right during gait, for example when we observe someone walking with a limp. For detailed analysis of gait using observational gait analysis, a systematic approach is required that considers the various components of gait at a level of detail that cannot be achieved by an overall visual impression. Clinical assessment using observational gait analysis therefore considers the movements and symmetry of movements of individual body segments and joints, observed using systematic methodologies.

The potential unreliability of observational gait analysis has been the subject of considerable research. It has been shown that the use of a systematic approach to observation and recording improves reliability (Krebs et al. 1985, Lord, Halligan, and Wade 1998, Perry 2002, Read et al. 2003, Toro, Nester, and Farren 2007a,b, Rathinam et al. 2014, Gor-García-Fogeda et al. 2016, University of Oklahoma Health Sciences Center 2002). An example of a systematic checklist for observational gait analysis is shown in Figure 3.6. Using such checklists results in good agreement between different observers when a binary scale is employed, such as when a movement pattern is noted as being "present" or "absent" in response to a specific question pertaining to the movement patterns, or when a simple three-point Likert scale, such as "present", "maybe present" or "absent", is used to assess the presence or absence of a movement or pattern of movement.

In 2016, the reliability and validity of assessing children with cerebral palsy using the Edinburgh Visual Gait Score were investigated (del Pilar Duque Orozco et al. 2016). It was found that levels of agreement between observers increased with experience of gait analysis, and that the assessment of the distal segments (foot, ankle and knee) were more reliable than the assessment of the proximal segments (hip, pelvis and trunk) (Figure 3.7). The observer's previous experience in gait analysis is an important factor affecting the reliability of observational gait analysis, and appropriate

		Stance Phase				Swing Phase		
		Initial Contact	Loading Response	Mid-Stance	Terminal Stance	Pre-Swing	Initial Swing	Mid-Swing	Terminal Swing
Trunk	forward lean								
	backward lean								
	lateral lean								
Pelvis	no forward rotation (R/L)								
	no contralateral drop (R/L)								
	hiking (R/L)								
Hip	inadequate extension								
	circumduction/abduction								
Knee	excessive flexion								
	uncontrolled extension								
	inadequate flexion								
Ankle/Foot	foot slap								
	forefoot contact								
	foot flat contact								
	late heel off								
	contralateral vaulting								

FIGURE 3.6 Gait Analysis Checklist. Adapted from Professional Staff Association of Rancho Los Amigos Medical Center (1989).

		Percentage of complete agreement		
		High level	Medium level	No-experience
Foot	Foot – initial contact	91	90	87
	Foot – heel lift	82	81	76
	Foot – max ankle dorsiflexion in stance	66	62	55
	Foot – hindfoot varus/valgus	71	72	65
	Foot – rotation	81	63	62
	Foot – clearance in swing	84	81	76
	Foot – max ankle dorsiflexion in swing	82	80	72
Knee	Progression angle	82	77	73
	Peak extension in stance	82	77	76
	Extension in terminal swing	70	70	55
	Peak flexion in swing	80	76	61
Hip	Peak extension in stance	78	78	75
	Peak flexion in swing	83	71	68
Pelvis	Obliquity at mid-stance	76	75	75
	Rotation at mid-stance	75	66	41
	Peak sagittal position	76	71	52
	Max lateral shift	80	65	63

FIGURE 3.7 Percentage of complete agreement between observers with different levels of experience. Adapted from del Pilar Duque Orozco (2016).

training is therefore required (del Pilar Duque Orozco et al. 2016). The importance of previous experience has been shown by a number of studies (Read et al. 2003, Toro, Nester, and Farren 2007b, Brown et al. 2008, Lu et al. 2015, Wellmon et al. 2015). Observational gait analysis has proved to be a very useful tool in the clinical assessment and classification of patients with a variety of pathological gait patterns, and has the potential to contribute to the assessment of non-pathological gait in individuals in a variety of situations.

VIDEO-BASED GAIT ANALYSIS

The use of video recording to examine gait has been widespread since the 1990s, and previous to that gait analysis was commonly undertaken using images captured on film. The use of a series of still images taken in rapid succession to analyse gait dates back to the work of Eadweard Muybridge in the late 1870s (Baker 2007). His pioneering work on locomotion used multiple cameras to create a motion-like image from individual photographs that were then viewed on what was termed a zoopraxiscope. Interestingly, this work was performed to settle a debate on whether all four legs of the horse were off the ground at the same time when they are running, which they are. In today's world, we take for granted the ability to establish such things by simply recording and then slowing down or pausing a video for more detailed analysis. In the 1990s, video analysis was enhanced greatly with the development of portable cameras and memory cards, which overcame a previous problem: a portable way to store the footage. The advantages of being able to video the gait of a subject include:

- reducing the number of repeated walks needed for analysis
- the ability to analyse events, positions and movements frame by frame and in slow motion
- the ability to compare gait over time and in different situations
- in some cases, the ability to gain objective data such as cadence, joint angles and timings of movements from the footage
- being able to share the footage with others and gain their opinions
- the production of a permanent record of a person's gait
- the use of footage as a tool when teaching or learning the art of gait analysis
- depending on the resolution of the footage, the ability to zoom in and study a joint or other feature of interest in detail

A variety of digital devices are now used to make recordings from which gait information can be gained, including cameras, phones and body-worn equipment. Important features to consider when selecting a device with which to capture footage for use in gait analysis include automatic focus, a zoom lens, the ability to capture data in low light, and a high-speed mode that eliminates blurring due to movement. Most digital video devices now capture images at 25 frames per second or more, with many smartphones now having frame rates of up to 60 frames per second. The more frames per second, the more information is recorded in a given time period. Many gait laboratories record video data directly into a computer, which can be synchronised with data collection from other motion analysis program systems, and which allows playback and analysis using freeze frame, frame by frame advance and variable slow speed play, which facilitate the observation of movements which are too fast for the unaided eye to determine.

In the laboratory or clinical context, it is important to consider both the positioning of the recording devices as well as how many cameras will be needed before the recording session begins. A number of studies have reported on the importance of the angle of the camera relative to the subject being recorded (Read et al. 2003, Toro, Nester, and Farren 2007b, Brown et al. 2008, Lu et al. 2015, Wellmon et al. 2015). The devices should be positioned to capture images perpendicular to the plane[3] of movement being assessed, which is dependent on the purpose of the analysis. Video recording captures three-dimensional movement, but only in two dimensions, and the position of the camera relative to the subject is therefore an important factor in any subsequent analysis. The 2018 study of Reinking et al. demonstrated higher inter-rater and intra-rater reliability of the assessment of sagittal plane movement compared to that frontal plane movement, using two-dimensional software (Reinking et al. 2018). Three-dimensional video analysis continues to be the "gold standard" for gait analysis due to the inherent limitations of analysing three-dimensional movement using two-dimensional image capture and software. Three-dimensional software programmes are available which are designed to take two-dimensional video from multiple cameras at different angles relative to the subject in order to perform more precise analysis (Maykut et al. 2015). Three-dimensional analysis using multiple video cameras commonly involves placing markers on key landmarks and videoing from multiple angles, and can therefore be time consuming to perform and analyse and inconvenient to carry out in the clinical context (Mukaino et al. 2018). However, there are systems that do not require the use of such markers and can therefore be used to produce three-dimensional information from multiple two-dimensional sources, without prior marking of the subject.

Although visual gait analysis using simple video recording is largely subjective, objective data can be derived using appropriate software (Reinking et al. 2018). Finkbiner et al. (2017) carried out a pilot study looking into the validity of using smartphones for analysing movement. The results of this study demonstrated an average of a two- to six-degree range of measurement error in joint angles, which was determined to be within a clinically acceptable range. While this study notes that three-dimensional motion analysis continues to be more precise, analysis of gait using in-app technology is a technique that can be easily implemented. Although the measurements produced are susceptible to error, as the movement at a joint may not be captured from an angle exactly perpendicular to the plane of motion, reasonable data regarding joint movements of the lower limb in the sagittal plane and some joint movements in the frontal plane can be obtained.

A usual routine for performing gait analysis using video recording is to ask the subject to wear shorts, a swim suit, or close-fitting clothing so that the masking effect of the clothing on the observation of movement and position is minimised. It is important that the subject should walk in their usual manner, so they should wear their usual footwear. The subject should complete a number of practice walks before starting video recording, to facilitate the adoption of usual gait in the unfamiliar situation. Two camera positions are commonly used, adjusted to show the whole body from head to feet, one viewing from the side of the walkway, capturing a sagittal plane view of the subject, and another from the end of the walkway, capturing a frontal plane view, as shown in Figure 3.8. The subject is recorded as they walk the length of the walkway, turn, and walk back to their starting position. In this way, footage is captured of both sides, front and back of the subject.

This process may be repeated with the cameras adjusted to show a close-up of the body from the waist down, or of the body segments of particular interest. It is often helpful to mark the subject's skin, where exposed, for example using an eyebrow pencil or whiteboard marker, to enhance the visibility of anatomical landmarks in the recording. More elaborate markers

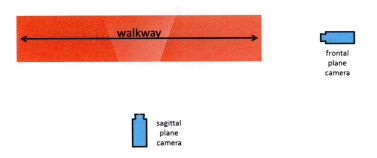

FIGURE 3.8 A typical camera set up for video analysis of gait.

can also be used to improve the accuracy of particular observations, such as the surface-mounted transverse plane rotation indicators fitted to subjects by Hillman et al. (1998). Subjects should not be able to see themselves while they are being videoed, as this presents a distraction to the subject and can alter their gait.

If possible, it is a good idea to review the recording before the subject leaves, in case the process needs to be repeated for any reason. The analysis is performed post recording, allowing time to be taken over the systematic observation of the features of gait.

THREE-DIMENSIONAL MOTION ANALYSIS

Although three-dimensional data regarding gait can be derived by using two-dimensional images from multiple video cameras, there are a number of methodologies that collect primary three-dimensional data. These include accelerometers, gyroscopes, the use of electromagnetic fields and optoelectronic stereophotogrammetry, the latter now being the most commonly used method of three-dimensional motion analysis. Optoelectronic stereophotogrammetry is used to triangulate the three-dimensional location of markers placed on the subject, using active or passive makers, emitting or reflecting infra-red light. Such systems produce real-time motion data relating to the position of each marker relative to the system, from which derived data can be calculated showing the relative movement of the markers. The use of multiple non-collinear markers attached to adjacent body segments allows the determination of intersegment positions, angles and accelerations. These systems can produce accurate and reliable three-dimensional motion analysis data, with low error rates and high levels of reliability. They are, however, relatively expensive, relatively complex to use and require the use of markers attached to the subject prior to data collection.

NOTES

1. Pronation of the foot is a triplanar movement in which the foot everts, dorsiflexes and abducts.
2. Supination of the foot is a triplanar movement in which the foot inverts, plantarflexes and adducts.
3. See Levine et al. 2012 for an explanation of body planes and their use in describing human movement (Levine, Richards, and Whittle 2012).

REFERENCES

Baker, Richard. 2007. "The history of gait analysis before the advent of modern computers." *Gait and Posture* 26(3):331–342.

Bresler, B., and J. P. Frankel. 1950. "The forces and moments in the leg during level walking." *American Society of Mechanical Engineers Transactions* 72:27–36.

Brown, C. R., S. J. Hillman, A. M. Richardson, J. L. Herman, and J. E. Robb. 2008. "Reliability and validity of the Visual Gait Assessment Scale for children with hemiplegic cerebral palsy when used by experienced and inexperienced observers." *Gait and Posture* 27(4):648–652. doi: 10.1016/j.gaitpost.2007.08.008.

Collado-Vázquez, S., and J. M. Carrillo. 2015. "Balzac and human gait analysis." *Neurología (English Edition)* 30(4):240–246.

del Pilar Duque Orozco, Maria, Oussama Abousamra, Chris Church, Nancy Lennon, John Henley, Kenneth J. Rogers, Julieanne P. Sees, Justin Connor, and Freeman Miller. 2016. "Reliability and validity of Edinburgh visual gait score as an evaluation tool for children with cerebral palsy." *Gait and Posture* 49:14–18. doi: 10.1016/j.gaitpost.2016.06.017.

Finkbiner, J. Monica, Kira M. Gaina, Marie C. McRandall, Megan M. Wolf, Vicky M. Pardo, Kristina Reid, Brian Adams, and Sujay S. Galen. 2017. "Video movement analysis using smartphones (ViMAS): A pilot study." *Journal of Visualized Experiments* (121):1–7. doi: 10.3791/54659.

Garrison, Fielding H. 1929. *An Introduction to the History of Medicine.* Philadelphia, PA: WB Saunders.

Gor-García-Fogeda, María Dolores, Roberto Cano de la Cuerda, María Carratalá Tejada, Isabel Mª Alguacil-Diego, and Francisco Molina-Rueda. 2016. "Observational gait assessments in people with neurological disorders: A systematic review." *Archives of Physical Medicine and Rehabilitation* 97(1):131–140. doi: 10.1016/j.apmr.2015.07.018.

Hillman, Susan J., M. Elizabeth Hazelwood, Ian R. Loudon, and James Robb. 1998. "Can transverse plane rotations be estimated from video tape gait analysis?" *Gait and Posture* 8(2):87–90. doi: 10.1016/S0966-6362(98)00028-9.

Inman, Verne T., Henry J. Ralston, and Frank Todd. 1981. *Human Walking.* Baltimore, MD: Williams & Wilkins.

References

Krebs, David E., Joan E. Edelstein, and Sidney Fishman. 1985. "Reliability of observational kinematic gait analysis." *Physical Therapy* 65(7):1027–1033. doi: 10.1093/ptj/65.7.1027.

Levine, David, Jim Richards, and Michael Whittle. 2012. *Whittle's Gait Analysis*. 5th edition. London, UK: Churchill Livingstone, Elsevier.

Lord, S. E., P. W. Halligan, and D. T. Wade. 1998. "Visual gait analysis: The development of a clinical assessment and scale." *Clinical Rehabilitation* 12(2):107–119.

Lu, Xi, Nan Hu, Siyu Deng, Jun Li, Shuyan Qi, and Sheng Bi. 2015. "The reliability, validity and correlation of two observational gait scales assessed by video tape for Chinese subjects with hemiplegia." *Journal of Physical Therapy Science* 27(12):3717–3721.

Mukaino, Masahiko, Kei Ohtsuka, Hiroki Tanikawa, Fumihiro Matsuda, Junya Yamada, Norhide Itoh, and Elichi Saitoh. 2018. "Clinical-oriented three-dimensional gait analysis method for evaluating gait disorder." *Journal of Visualized Experiments* (133):1–7. doi: 10.3791/57063.

Maykut, Jennifer N., Jeffery A. Taylor-Haas, Mark V. Paterno, Christopher A. DiCesare, and Kevin R. Ford. 2015. "Concurrent validity and reliability of 2d kinematic analysis of frontal plane motion during running." *International Journal of Sports Physical Therapy* 10(2):136–146.

Murray, Melissa P. 1967. "Gait as a total pattern of movement." *American Journal of Physical Medicine* 46(1):290–333.

Murray, Melissa P., A. Bernard Drought, and Ross C. Kory. 1964. "Walking patterns of normal men." *Journal of Bone and Joint Surgery* 46A:335–360.

Murray, Melissa P., Rory C. Kory, and Susan B. Sepic. 1970. "Walking patterns of normal women." *Archives of Physical Medicine and Rehabilitation* 51(11):637–650.

Nene, Anand, Ruth Mayagoita, and Peter Veltink. 1999. "Assessment of rectus femoris function during initial swing phase." *Gait and Posture* 9(1):1–9. doi: 10.1016/S0966-6362(98)00042-3.

Perry, Jacquelin. 1974. "Kinesiology of lower extremity bracing." *Clinical Orthopaedics and Related Research* 102:18–31. doi: 10.1097/00003086-197407000-00004.

Perry, Jacquelin. 2002. *Gait Analysis*. Vol. revised edtion (2001), *Normal and Pathological Function*. Downey, CA: Los Amigos Research & Education Center. Original edition (1992).

Professional Staff Association of Rancho Los Amigos Medical Center. 1989. *Observational Gait Analysis Handbook*. Downey, CA.

Rathinam, Chandrasekar, Andrew Bateman, Janet Peirson, and Jane Skinner. 2014. "Observational gait assessment tools in paediatrics – A systematic review." *Gait and Posture* 40(2):279–285. doi: 10.1016/j.gaitpost.2014.04.187.

Read, Heather S., M. Elizabeth Hazlewood, Susan J. Hillman, Robin J. Prescott, and James E. Robb. 2003. "Edinburgh visual gait score for use in cerebral palsy." *Journal of Pediatric Orthopaedics* 23(3):296–301.

Reinking, Mark F., Leigh Dugan, Nolan Ripple, Karen Schleper, Henry Scholz, Jesse Spadino, Cameron Stahl, and Thomas G. McPoil. 2018. "Reliability of two-dimensional video-based running gait analysis." *International Journal of Sports Physical Therapy* 13(3):453–461.

Rigas, Costantinos. 1984. "Spatial parameters of gait related to the position of the foot on the ground." *Prosthetics and Orthotics International* 8(3):130–134.

Rose, Jessica, and James G. Gamble. 1994. *Human Walking.* 2nd edition. Baltimore, MD: Williams & Watkins.

Steindler, Arthur. 1953. "A historical review of the studies and investigations made in relation to human gait". *Journal of Bone and Joint Surgery* 35:540–542.

Sutherland, D. H. 2001. "The evolution of clinical gait analysis part I: Kinesiological Emg." [In English]. *Gait and Posture* 14(1):61–70. doi: 10.1016/S0966-6362(01)00100-X.

Sutherland, D. H. 2002. "The evolution of clinical gait analysis. Part II: Kinematics." *Gait Posture* 16:159–179.

Sutherland, D. H. 2005. "The evolution of clinical gait analysis. Part II: Kinetics and energy assessment." *Gait Posture* 21:447–461.

Toro, Brigitte, Christopher J. Nester, and Pauline C. Farren. 2007a. "The development and validity of the Salford Gait Tool: An observation-based clinical gait assessment tool." *Archives of Physical Medicine and Rehabilitation* 88(3):321–327.

Toro, Brigitte, Christopher J. Nester, and Pauline C. Farren. 2007b. "Inter- and intraobserver repeatability of the Salford Gait Tool: An observation-based clinical gait assessment tool." *Archives of Physical Medicine and Rehabilitation* 88(3):328–332.

Turnbull, George I., and James C. Wall. 1985. "The development of a system for the clinical assessment of gait following a stroke." *Physiotherapy* 71:294–298.

University of Oklahoma Health Sciences Center. 2002. "Observational gait analysis." https://ouhsc.edu/bserdac/dthompso/web/gait/knmatics/oga.htm. Accessed February 24, 2019.

Wellmon, Robert, Amy Degano, Joseph A. Rubertone, Sandra Campbell, and Kelly A. Russo. 2015. "Interrater and intrarater reliability and minimal detectable change of the Wisconsin Gait Scale when used to examine videotaped gait in individuals post-stroke." *Archives of Physiotherapy* 5(1):11.

The legal context of forensic gait analysis

PART 1: THE LEGAL CONTEXT IN NORTH AMERICA
Emma Cunliffe

Legal rules are important at two phases of the work performed by a forensic gait analyst. During the *investigative phase*, their work will be regulated by rules and procedures about matters such as what information may be collected by police, how that information can be shared with forensic analysts who are not State employees, and what information must be disclosed to a criminal suspect. A forensic gait analyst should make and keep full records of all information received (including physical exhibits, correspondence, and even conversations with investigating officers and prosecutors), procedures undertaken, and results obtained. Upon request, they must provide this information to the police or prosecutor. Notably, if and when a suspect is charged, the suspect will likely be entitled to receive copies of all of this information. Accordingly, it is important for a forensic gait analyst both to prepare and retain good records and to ensure that they are comfortable explaining their practices and procedures to a court if questions eventually arise. A useful resource when preparing for cross-examination about these matters is Gary Edmond and co-authors' "Model forensic science" (Edmond et al. 2016).

In many jurisdictions, quality assurance and quality management regulations will also govern the work performed during the investigative phase (Doyle 2018). These regulations may specify minimum qualifications, record-keeping responsibilities, error identification and correction processes, and mandatory procedures. Further obligations may be imposed by the forensic practitioner's professional association. When the forensic practitioner is a member of a professional body concerned with forensic practice and human identification (e.g. the International Association for Identification, or the American Academy of Forensic Sciences), that association will also have rules that govern members' work during the investigative phase. Before accepting forensic work, it is important for a practitioner to ensure that they are familiar with these requirements in their own jurisdiction and in the jurisdiction in which the case is being investigated.

During the *evaluative phase*, a forensic practitioner's work is presented to a court and evaluated through an adversarial process in which both the prosecutor and the accused person are entitled to inquire into the forensic practitioner's work. Much of the remainder of this section focusses on the legal rules and regulations that govern a forensic practitioner's work in the evaluative phase. However, there is also obvious continuity between the work performed during the investigative phase and that performed during the evaluative phase. A forensic practitioner in the investigative phase should prepare their work in the expectation that they will be asked to explain and defend their procedures and conclusions during the evaluative phase.

PRINCIPLES OF THE LAW AND THE LEGAL SYSTEM

The United States of America and Canada have legal systems based on (originally English) common law principles, but with the added dimension of constitutionalised human rights protections that regulate criminal investigations, police behaviour, and criminal trial practice. In the US, the power to enact criminal law is assigned to both the federal government and individual states. In Canada, criminal law is allocated exclusively to the federal government, but police and courts administration are provincial responsibilities. These features mean that a patchwork of legal rules and processes exists across the North American continent. These rules and processes vary quite widely, and so this guide addresses the largest jurisdictions (US and Canadian federal rules) and general principles.

In the US, the federal government and each state maintains its own courts, police agencies, rules of evidence, and criminal legislation. However, constitutional rules are universal – including the principles that an accused person has a right to a lawyer, to a trial by jury, and to question witnesses, including expert witnesses, who give testimony against them. Similarly, it is a constitutional principle that the police must obtain a warrant from a judge before securing some kinds of evidence, especially before an arrest is made (Legal Information Institute n.d.). In the Federal US jurisdiction, the rules that govern much of the investigation and trial process are the *Federal Rules of Criminal Procedure* and the *Federal Rules of Evidence* (Legal Information Institute 2015, 2018). In Canada, the courts that hear criminal matters in the first instance are provincial courts, though in more serious cases the judge will be federally appointed. Again, constitutional rules that govern matters such as what information must be disclosed to a suspect, an accused person's right to challenge evidence, and the right to a fair trial apply universally across all Canadian courts.

Forensic gait analysis is treated as expert opinion evidence in the US and Canada, which means that it will be subject to judicial evaluation before a forensic practitioner will be permitted to give evidence. Article VII of the US *Federal Rules of Evidence* regulates opinion and expert evidence. Rule 702 provides:

> A witness who is qualified as an expert by knowledge, skill, experience, training, or education may testify in the form of an opinion or otherwise if:
>
> (a) the expert's scientific, technical, or other specialized knowledge will help the trier of fact to understand the evidence or to determine a fact in issue
> (b) the testimony is based on sufficient facts or data
> (c) the testimony is the product of reliable principles and methods
> (d) the expert has reliably applied the principles and methods to the facts of the case

All criteria must be satisfied in order for the evidence to be received by a Federal Court. Rule 702 and the cases that have interpreted it establish that the trial judge acts as a 'gatekeeper' to the admission of expert evidence (*Daubert v Merrell Dow Pharmaceuticals* 1993). This section supplies a relatively brief and general overview of Rule 702 and its associated caselaw. For those who are interested in learning more, a good source is Edward J. Imwinkelreid's "Regulating expert evidence in US courts: Measuring *Daubert's* impact" (2018).

As Imwinkelreid's title suggests, the leading decision is *Daubert v Merrell Dow Pharmaceuticals* (*Daubert v Merrell Dow Pharmaceuticals* 1993). In that case, a majority of the US Supreme Court held that the phrase 'scientific ... knowledge' in Rule 702 requires a trial judge to make 'a preliminary assessment of whether the reasoning or methodology underlying the testimony is scientifically valid and of whether that reasoning or methodology properly can be applied to' the issues in the case. For example, in a case in which a forensic practitioner relies on their training to claim that there is correspondence between features of gait shown in CCTV footage of a true perpetrator and features shown in covertly obtained footage of a police suspect, the judge must consider whether the methodology used by the forensic practitioner is scientifically valid and can properly be applied to the evidence to assist in determining the identity of the true perpetrator.

The US Supreme Court suggested that 'there is no definitive checklist' for determining scientific validity, but relevant considerations include:

(a) whether the technique can be tested (in circumstances where the true state of affairs is known to the person who designs the test) and, if so, whether it has been tested (Kaye 2016)
(b) whether the technique has been subjected to peer review and publication
(c) the known or potential rate of error
(d) whether the technique is generally accepted within a relevant scientific community

In a subsequent decision, the US Supreme Court clarified that, when presented with expert evidence, a trial judge should focus upon whether the specific technique used by the expert in this case has been validated (and not upon whether the field as a whole is regarded as valid) (*General Electric Co v Joiner* 1997). The Court also confirmed that where the nature of the expert evidence is technical or otherwise specialised (i.e. expert, but not scientific), the *Daubert* approach continues to apply with appropriate modifications to reflect the nature of validity or reliability within the relevant field. In short, whether expert testimony is scientific, technical, or specialised, 'the trial judge must determine whether the testimony has "a reliable basis in the knowledge and experience of [the relevant] discipline"' (*Kumho Tire Co v Carmichael* 1999).

Rule 702 places an onus upon the party who has commissioned the forensic gait analysis to establish that it is sufficiently valid to be admitted into the courtroom. *Daubert* and related cases explain what kind of information a court will need in order to make that determination. We can therefore expect that when forensic gait analysis is offered in a US Courtroom, it will be subject to the careful judicial scrutiny anticipated by the US Supreme Court. In this context, one may reasonably anticipate that the findings of the Royal Society and the Royal Society of Edinburgh and of some academic researchers about the empirical basis for forensic gait analysis are likely to prove significant to US courts (Edmond and Cunliffe 2016, The Royal Society and the Royal Society of Edinburgh 2017).

A forensic practitioner who is commissioned to prepare an opinion for US courts should also familiarise themselves with the 2009 National Research Council (NRC) report, *Strengthening Forensic Science in the US: A Path Forward* (National Research Council of the National Academies of Science 2009) and the 2016 report of the President's Council of Advisors on Science and Technology (PCAST), *Forensic Science in Criminal Courts: Ensuring Scientific Validity of Feature-Comparison Methods* (2016).

In Canada, the admissibility of expert opinion evidence is governed by a mix of common law (i.e. judge made law) and statutory law. These rules are mostly federal. Section 657.3 of the *Criminal Code* requires that a party who wishes to rely on expert evidence must provide notice to the other party. This notice must include information such as the name of the expert witness, the areas in which they are qualified to testify, and the nature of their qualifications. If the expert witness will be called by the prosecutor, a copy of the expert report or a summary of the evidence they will give must also be supplied before the trial begins. (For an expert witness called by the accused person, that information must be supplied later in the proceeding. As a matter of practical consideration, this means that if a forensic practitioner is called by the prosecutor, they will not necessarily know what opinion an expert witness called by the defence will give.)

Canadian law also regulates the reliability and scientific validity of expert opinion evidence. In the leading case of *White Burgess Langille Inman v Abbott & Haliburton*, the Supreme Court of Canada held that a trial judge must conduct a two-stage inquiry. At the first stage, the trial judge will consider whether the expert opinion is relevant to an issue in the case, necessary to assist the fact-finder, and otherwise permitted by the rules of evidence; and whether the expert is properly qualified to provide the opinion. At the second, more searching, stage, the trial judge will closely assess the reliability and independence of the expert opinion evidence, weighing its value to the fact-finder against the risks presented by, for example, concerns about cognitive bias or shortcomings in the empirical research that underpins an expert's methods. The factors set out in *Daubert v Merrell Dow Pharmaceuticals* have also been applied by Canadian courts (*R v J(JL)* 2000). For those who are interested in learning more about the Canadian approach, a comprehensive discussion is provided in Emma Cunliffe's "A new Canadian paradigm? Judicial gatekeeping and the reliability of expert evidence" (Cunliffe 2018).

The Supreme Court of Canada has, in recent years, placed growing emphasis on the trial judge's responsibility to safeguard the reliability of expert testimony, both at the time of deciding whether to admit an expert's evidence and throughout their testimony (*R v Sekhon* 2014). This means that a forensic practitioner may expect a searching evaluation of the basis for their opinion before they are permitted to testify, and that they may also be reminded of the proper scope of their evidence while they testify.

In Canada, forensic gait analysis was admitted to a trial court in *R v Aitken* (2008). The BC Court of Appeal concluded that the forensic gait analysis was admissible in the circumstances of this case (*R v Aitken* 2012). This decision has been criticised (Cunliffe and Edmond 2013). Since *Aitken* was decided, the Supreme Court of Canada has clarified (and arguably strengthened) the rules that govern the admissibility of expert opinion evidence (*R v Sekhon* 2014). Additionally, in Canada the admissibility of expert opinion evidence is decided on a case-by-case basis, so a prior decision to admit a given kind of expert opinion evidence does not bind the trial judge (*R v Trochym* 2007). For these reasons, a forensic practitioner who is preparing to testify to forensic gait analysis in a Canadian court should anticipate that their evidence is likely to be more rigorously scrutinised than occurred in *R v Aitken*. In addition to reviewing the NRC and PCAST reports recommended above, a forensic practitioner who is preparing to testify in a Canadian court should read the most relevant chapters of Stephen T. Goudge's report on the reliability of expert testimony in criminal proceedings (2008).

ETHICAL ROLE AND RESPONSIBILITIES OF THE EXPERT WITNESS

Forensic practitioners must also have a strong appreciation of the ethical role and responsibilities of the expert witness. In the USA and Canada, a sobering number of wrongful convictions have been linked to erroneous and misleading forensic science evidence (Garrett and Neufeld 2009, Cunliffe and Edmond 2017). Criminal charges in both jurisdictions must be proven by the prosecutor 'beyond a reasonable doubt', and forensic science is frequently significant to the discharge of this responsibility. Studies suggest that overconfidence and over claiming are more frequent causes of error in forensic science testimony than fraud or malicious misconduct. The overwhelming lessons that emerge from these studies and associated government reports are that a forensic practitioner must familiarise themselves with the research on the scientific validity of the techniques they are using, scrupulously avoid overstating the state of research in their field, and avoid expressing personal views about the results they have obtained even if their techniques have been validated. Statements that suggest that there is no possibility of error are always improper.

Many professional associations and the International Association for Identification have their own rules with regard to the ethical responsibilities of their members. In some jurisdictions, courts will have specific rules and may ask an expert witness to sign a declaration affirming their understanding of and willingness to abide by these rules (e.g. *Ontario Rules of Civil Procedure* 1990). At the time of writing this edition, the US Organization of Scientific Area Committees for the Forensic Sciences was in the process of preparing a Code of Practice for Forensic Scientists. Before testifying, a forensic practitioner should identify which codes apply to their work and ensure that they are familiar with these rules.

Beyond these codes of conduct, some general suggestions may assist:

> the expert witness is not simply representing herself but the knowledge her field has about some topic.
>
> **Sanders 2007**

This means that an expert witness should regard their task as being to convey the state of knowledge in their field, including any controversy or debates that may be important to the court's task in a given case. If they take a position that is not shared by others in their field, the nature of the disagreement should be explained, along with their reasons for adopting the position they prefer.

> An independent expert is not affected by the goals of the party for which she was retained, and is not reticent to arrive at an opinion that fails to support the client's legal position.
>
> **Lubet 1998**

An expert witness must provide independent and impartial advice, in a manner that is not influenced by the party who called them. It is therefore important to ensure that any assumptions underlying the opinion, limits to that opinion, and arguments against the correctness of the opinion are conveyed to the court.

In Canada, these principles were adopted by the Supreme Court of Canada in *White Burgess Langille Inman v Abbott & Haliburton* (2015). In that case, Justice Cromwell explained that the 'acid test is whether the expert's opinion would not change regardless of which party retained him or her'. An expert who is biased, or otherwise unable to discharge the ethical duty to remain impartial and provide objective assistance to the court, may well be excluded.

Acting as an expert witness is a solemn responsibility. By participating in this way in the criminal legal system, one is committing oneself to a fearless search for the truth and a repugnance for wrongful convictions. For this reason, a prospective expert witness should ensure that they allow themselves plenty of time to prepare their testimony, and they must always be scrupulous in the discharge of their professional responsibilities.

REFERENCES

Cunliffe, Emma, and Gary Edmond. 2013. "Gaitkeeping in Canada: Missteps in Assessing the Reliability of Expert Testimony." *Canadian Bar Review* 92:327–368.

Cunliffe, Emma, and Gary Edmond. 2017. "What Have We Learned? Lessons from Wrongful Convictions in Canada." In: Berger, B., E. Cunliffe, and J. Stribopoulos (eds.) *To Ensure that Justice Is Done: Essays in Memory of Marc Rosenberg*. Toronto: Thomson Reuters.

Cunliffe, Emma. 2018. "A New Canadian Paradigm? Judicial Gatekeeping and the Reliability of Expert Evidence." In: Roberts, P., and M. Stockdale (eds.) *Forensic Science Evidence and Expert Witness Testimony*. Cheltenham: Edward Elgar Publishing.

Daubert v. Merrell Dow Pharmaceuticals, Inc., 509 U.S. 579 (1993).

Doyle, Sean. 2018. *Quality Management in Forensic Science*. London: Academic Press.

Edmond, Gary, and Emma Cunliffe. 2016. "Cinderella Story: The Social Production of a Forensic Science." *Journal of Criminal Law and Criminology* 106:219.

Edmond, G., B. Found, K. Martire, K. Ballantyne, D. Hamer, R. Searston, M. Thompson, E. Cunliffe, R. Kemp, M. San Roque, J. Tangen, R. Dioso-Villa, A. Ligertwood, D. Hibbert, D. White, G. Ribeiro, G. Porter, A. Towler, and A. Roberts. 2016. "Model Forensic Science." *Australian Journal of Forensic Sciences* 48(5):496–537.

Garrett, Brandon, and Peter Neufeld. 2009. "Invalid Forensic Science Testimony and Wrongful Convictions." *Virginia Law Review* 95:1–97.

General Electric Co v Joiner, 522 US 136 (1997).

Goudge, Stephen. 2008. *Inquiry into Pediatric Forensic Pathology in Ontario*. Inquiry into Pediatric Forensic Pathology in Ontario. Ontario: The Ministry of the Attorney General.

Imwinkelreid, Edward J. 2018. "Regulating Expert Evidence in US Courts: Measuring Daubert's Impact." In: Roberts, P., and M. Stockdale (eds.) *Forensic Science Evidence and Expert Witness Testimony: Reliability Through Reform?* Cheltenham: Edward Elgar Publishing.

Kaye, David H. 2016. "Hypothesis Testing in Law and Forensic Science: A Memorandum." *Harvard Law Review F* 130:127.

Kumho Tire Co v Carmichael, 526 US 137 (1999).

Legal Information Institute. 2015. *Federal Rules of Evidence*. Accessed 01/06/19. https://www. law.cornell.edu/rules/fre.

Legal Information Institute. 2018. *Federal Rules of Criminal Procedure*. Accessed 01/06/19. https://www.law.cornell.edu/rules/frcrmp.

Legal Information Institute. n.d. *Criminal Procedure*. Accessed 01/06/2019. https://www.law. cornell.edu/wex/criminal_procedure.

Lubet, Steven. 1998. "Expert Witnesses: Ethics and Professionalism." *Georgetown Journal of Legal Ethics* 12:465.

National Research Council. 2009. *Strengthening Forensic Science in the United States: A Path Forward*. Washington, DC: National Academies Press.

Ontario Rules of Civil Procedure. Form. 53. R.R.O. 1990, Reg 194.

R v Aitken, 2008 BCSC 1423 (BC Supreme Court); Decision to Admit Evidence Upheld in *R v Aitken*, 2012 BCCA 134 (BC Court of Appeal).

R v Aitken, 2012 BCCA 134 (BC Court of Appeal).

R v J(JL), 2000 SCC 51.

R v Sekhon, 2014 SCC 15.

R v Trochym, 2007 SCC 6.

Sanders, Joseph. 2007. "Expert Witness Ethics." *Fordham Law Review* 76:1539.

The President's Council of Advisors on Science and Technology. 2016. *Report to the President: Forensic Science in Criminal Courts: Ensuring Scientific Validity of Feature-Comparison Methods*. Washington: President's Council of Advisors on Science and Technology.

The Royal Society and the Royal Society of Edinburgh. 2017. *Forensic Gait Analysis: A Primer for Courts*. London: The Royal Society.

White Burgess Langille Inman v Abbott & Haliburton, 2015 SCC 23.

PART 2: THE LEGAL CONTEXT IN THE UNITED KINGDOM

Graham Jackson

The rules, requirements and procedures in relation to expert evidence in the jurisdictions of the United Kingdom reflect to a large extent the principles and practices described for the jurisdictions of North America. While there are three different jurisdictions within the UK – England and Wales, Scotland and Northern Ireland – there is a degree of overlap and commonality between the laws and practices in these jurisdictions. Arguably, the jurisdiction of England and Wales has the most developed framework for regulating and admitting expert evidence.

The main sources of guidance include:

1. *Criminal Practice Directions, Criminal Procedure Rules* (for England and Wales) 2015
2. Crown Prosecution Service (of England and Wales) *Expert Evidence Legal Guidance 2019*
3. Forensic Science Regulator *Codes of Practice and Conduct* 2017 and *Legal Obligations* 2019 and other documents published by the UK Forensic Science Regulator
4. applicable ISO and UKAS standards
5. regulations of professional bodies
6. stated cases from the Courts of Appeal

These sources provide in-depth, comprehensive guidance on the behaviour and practices required of an expert witness. Inevitably there is repetition between these sources and we highlight sections from just two of them. Firstly, paragraph 1A.3 of the *Criminal Practice Directions* (2015) states:

> The Criminal Procedure Rules and the Criminal Practice Directions are the law. Together they provide a code of current practice that is binding on the courts to which they are directed, and which promotes the consistent administration of justice. Participants must comply with the Rules and Practice Direction, and directions made by the court, and so it is the responsibility of the courts and those who participate in cases to be familiar with, and to ensure that these provisions are complied with.

It is therefore of vital importance for all experts providing evidence in England and Wales that they are thoroughly familiar with the requirements of the *Criminal Practice Directions/Criminal Procedure Rules*. Part 19.2 states:

(1) An expert must help the court to achieve the overriding objective—
   (a) by giving opinion which is—
      (i) objective and unbiased, and
      (ii) within the expert's area or areas of expertise; and
   (b) by actively assisting the court in fulfilling its duty of case management under rule 3.2, in particular by—
      (i) complying with directions made by the court, and
      (ii) at once informing the court of any significant failure (by the expert or another) to take any step required by such a direction.
(2) This duty overrides any obligation to the person from whom the expert receives instructions or by whom the expert is paid.
(3) This duty includes obligations—
   (a) to define the expert's area or areas of expertise—
      (i) in the expert's report, and
      (ii) when giving evidence in person;
   (b) when giving evidence in person, to draw the court's attention to any question to which the answer would be outside the expert's area or areas of expertise;
   (c) to inform all parties and the court if the expert's opinion changes from that contained in a report served as evidence or given in a statement; and
   (d) to disclose to the party for whom the expert's evidence is commissioned anything—
      (i) of which the expert is aware, and
      (ii) of which that party, if aware of it, would be required to give notice under rule 19.3(3)(c).

Secondly, the *Codes of Practice and Conduct* published by the Forensic Science Regulator (Issue 4 October 2017),[3] while having no formal force in law, are nevertheless viewed as a requirement on all forensic science providers. To quote the Regulator in her "Foreword" to the *Codes*:

> I strongly urge the many organisations that are not yet compliant with the required standards to prioritise quality: it cannot be regarded as a poor second to operational delivery. Whilst the standards are not yet mandated by law, compliance is not optional.
>   In my Annual Report published January 2017, I stated clearly that failure to comply with the Regulator's standards needed to be disclosed as this could significantly detract from the credibility of a forensic science practitioner and have a bearing on reliability. Therefore, I have revised the Code of Conduct to ensure it is sufficiently robust, requiring the highest standards of personal conduct and organisational compliance with quality standards. Individuals reporting scientific or technical work to the courts (whether called by prosecution or defence)

must now declare compliance with this Code of Conduct, and wording is provided to assist experts in fulfilling their obligations under the revised Criminal Practice Directions.

Practitioners in forensic gait analysis in the UK are subject to the same legal and professional requirements as are practitioners in the larger, main-stream forensic science providers. In addition they are also subject to the requirements of the *Code of practice for forensic gait analysis* (2019). It is necessary for all practitioners and their employing organisations not only to be familiar with the requirements but to demonstrate compliance with them.

REFERENCES

Chartered Society of Forensic Sciences and College of Podiatry in association with the Forensic Science Regulator. 2019. *Code of practice for forensic gait analysis*, Issue 1. Birmingham: The Forensic Science Regulator.

The Criminal Procedure Rules. The Criminal Practice Directions. October 2015 edition. Amended April, October and November 2016; February, April, August, October and November 2017; April and October 2018; and April 2019. Accessed July 2019. www.justice.gov.uk/courts/procedure-rules/criminal/rulesmenu-2015

Crown Prosecution Service (of England and Wales). *Expert Evidence Legal Guidance 2019*. Accessed July 2019. www.cps.gov.uk/legal-guidance/expert-evidence

Forensic Science Regulator. 2017. *Codes of Practice and Conduct: for Forensic Science Providers and Practitioners in the Criminal Justice System*, Issue 4. Birmingham: The Forensic Science Regulator.

Forensic Science Regulator. 2019. *Legal Obligations*, Issue 7. Birmingham: The Forensic Science Regulator.

5

The development of the forensic gait analysis quality assurance process in the UK

Sarah Reel

IDENTIFYING THE REQUIREMENT FOR QUALITY ASSURANCE IN FORENSIC SCIENCE

The majority of the forensic identification sciences, including those that examine fingerprints, voiceprints, footprints, shoeprints, bite marks, tool marks, firearm marks and forensic gait analysis, have been criticised in the past as lacking a scientific foundation (Pretty 2006, Cole 2007, Saks and

Faigman 2008, Saks and Koehler 2008) and have been referred to as the 'non-science forensic sciences' (Saks and Faigman 2008, p.150). The United States National Research Council report *Strengthening forensic science in the United States: A path forward* (National Research Council 2009) was particularly critical of a notable lack of a quality framework in the forensic science disciplines in terms of standards of proficiency, accreditation, standards of practice, alignment with recognised international standards, codes of ethics and laboratory-related quality assurances. At that time, the United States was not alone in reporting a lack of quality assurance in forensic science. The United Kingdom House of Commons Science and Technology Committee report, examining the state of forensic science in 2005, recommended an independent forensic science regulator be appointed to oversee quality regulation in recognition of similar flaws (2005). In response to this recommendation, the role of a facilitator for ameliorating the issue of poor standards within the forensic science disciplines in the UK was created in 2007 by the Home Secretary in the form of the Forensic Science Regulator, although no statutory powers were associated with the role. The Forensic Science Regulator, sponsored by the UK Home Office, was tasked with 'setting and monitoring quality standards for the use of forensic science in the Criminal Justice System' (The Law Commission 2009).

RECOGNITION OF REQUIREMENTS FOR QUALITY ASSURANCE IN FORENSIC GAIT ANALYSIS

The enormous task of providing standards regulations for every forensic science discipline in England and Wales by a Forensic Science Regulator devoid of statutory powers was never going to be completed quickly or easily. It was inevitable that a practitioner wishing to resist the development of standards could take advantage of a lack of statutory regulation. However, many practitioners envisaged that the lack of quality provision would have detrimental consequences to the criminal justice system as a whole, facilitating miscarriages of justice. Among these like-minded providers was a group of forensic podiatrists led by Professor Wesley Vernon OBE, a researcher and practitioner in this niche field, who recognised the need for a quality framework for their own discipline. The discipline of forensic podiatry includes the areas of identification of a person through podiatry records, footprint examination as an aid to identification, 'feet-in-shoes' examination and forensic gait analysis (Vernon 2006, DiMaggio and Vernon 2017). Although it is now recognised that appropriately qualified members of other professions could legitimately practice in the latter three of these areas of forensic practice, at the time they were often regarded as the domain of podiatrists. In 2006, just prior to the appointment of the Forensic Science Regulator, the Council for the Registration of Forensic Practitioners worked alongside this group in the UK to develop quality frameworks for forensic practitioners. The intention was to benchmark the quality of practicing forensic podiatrists' work, thus ensuring that their standards of practice adhered to 'a broad generic framework used across all registered specialities' (DiMaggio and Vernon 2017, p.244). In 2009, the Council for the Registration of Forensic Practitioners closed. However, the momentum to develop a quality framework for forensic podiatry continued. In the same year as the Council for the Registration of Forensic Practitioners' closure, the International Association for Identification published the *Forensic Podiatry Role and Scope of Practice* document on their website, including a section on forensic gait analysis (Vernon et al. 2010). This document was developed in response to concerns that some clinical podiatrists, with little or no forensic science training, may have been practising inappropriately in the forensic arena. These practitioners were aware of and adhered to clinical podiatry standards and regulations relevant to their clinical work, however, such standards were not necessarily relevant to a forensic science quality framework. The *Forensic Podiatry Role and Scope of Practice* document outlined the recommended training in order to practice in the discipline and the boundaries within which the forensic podiatrist should operate (DiMaggio and Vernon 2017). It also provided reassurance to others as to what the discipline covered and what tasks were outside the role and scope of forensic podiatry practice. What was written initially, in anticipation of the potential for problems to arise due to the lack of a framework and related concerns raised by the International Association for Identification Board of Directors became a key starting point in providing a quality framework for forensic gait analysis. Four months after the publication of the *Forensic Podiatry Role and Scope of Practice* document, the aforementioned National Research Council report (2009) was published in the United States with its recommendations regarding quality assurance within the forensic sciences. It was welcomed by the authors of the *Role and Scope* document as it supported the document's main objectives. However, it was clear that more work was necessary in terms of developing standards relating to competency testing, method validation, laboratory accreditation, research provision (for example in human observer bias) and quality control procedures and assurances (National Research Council 2009, p.19–26).

After the closure of the Council for the Registration of Forensic Practitioners, the group of forensic podiatrists looked for a suitable substitute for the continuation of quality development within their profession and began discussions with the Forensic Science Society (latterly the Chartered Society of Forensic Sciences), identifying assessment of professional competence as a key objective. The Society's Standards Committee, the Forensic Science Regulator and the University of Staffordshire (Gwinnett, Cassella, and Allen 2011) supported the group in developing competency tests to ensure the quality of forensic podiatry expert evidence by a provider was of an acceptable standard. This was intended to provide an affordable regulation option which set out to test the professional competence of the practitioner in this field. The first sets of assessments were carried out in September 2010 in footprint and footwear analysis, and separately in forensic gait analysis.

Certificates of Professional Competence were issued to successful candidates who had passed multiple choice assessments in both 'general' forensics and their specialism, plus a three-hour practical examination (Chartered Society of Forensic Sciences 2019a). The Society has since presented this model to the wider community and it is now offered to practitioners in the area of crime scene investigation. It is also to be offered to collision investigation and firearms investigation (Bellamy 2018). The certificate illustrates a practitioner's ongoing competence and commitment to continuing professional development. However, participation in the Certificate of Professional Competence scheme, like its original Council for the Registration of Forensic Practitioners counterpart, is voluntary. McCartney and Amoako (2018) argue the demise of the Council for the Registration of Forensic Practitioners centred on the effects of a 'self-selective system of individual accreditation' (p.3) which the authors suggested ultimately forced financial difficulties upon the organisation due to a lack of interest in the scheme. It is hoped that the scheme offered by the Chartered Society of Forensic Sciences will not incur the same fate.

In 2011, the UK Law Commission report acknowledged the suggestion by the UK Accreditation Service that 'judges should take account of the increased confidence that can be derived from the fact that an expert works within the context of an accredited organisation, which is regularly assessed by an independent, impartial national accreditation body' (the Law Commission 2011, p.7). Additionally, Skills for Justice (2014) noted that ongoing audits of competence in the workplace of an expert witness would ensure quality. Although this was acknowledged and understood by forensic podiatry practitioners, most individuals usually worked as sole traders from home-based facilities or from a small office at their usual work-place, as forensic gait analysis and forensic podiatry casework were infrequently undertaken. Few individuals at that time were working for what could be described as an 'organisation'. As a result, this type of quality assurance in the forensic workplace seemed largely irrelevant, many regarding such accreditation as unnecessary.

CREATION OF A WRITING GROUP FOR FORENSIC PODIATRY STANDARDS

Also in 2011, Version 1 of *The Codes of Practice and Conduct for Forensic Science Providers and Practitioners in the Criminal Justice System* was published by the Office of the Forensic Science Regulator, Andrew Rennison (Forensic Science Regulator 2011). The *Codes* were strongly aligned with the International Standard BS EN ISO/IEC 17025:2005 for testing and calibration laboratories, but once again sections of the document did not seem relevant for forensic podiatrists including forensic gait analysts. Although the group was keen to provide exemplary quality assurance, they could not engage fully with a set of codes of practice and conduct that seemed to lie outside their environment. A few members of the group, including Wesley Vernon, met with Andrew Rennison to discuss the perceived limitations of the published codes and a potential way forward. The problems relating to relevancy encountered by forensic podiatry providers had also been shared by other professions including anthropology, archaeology, vehicle investigation, presumptive drug testing, finger mark comparison, digital data analysis and fire investigation (Rennison 2012). The Forensic Science Regulator suggested a separate 'Code of Practice, Ethics and Professional Standards in Forensic Podiatry' should be developed using similar, previously submitted forensic anthropology and archaeology draft documents as a template. The group worked together on behalf of the Society of Chiropodists and Podiatrists' (now the College of Podiatry) Forensic Podiatry Special Interest Group to create a first draft outlining duties and responsibilities, ethics and conduct, professional standards and competence including training and education, standard operating procedures, the expert witness report and attendance at court for the four areas of forensic podiatry including forensic gait analysis. Using headings taken from the anthropology and archaeology documents, the group embarked on a mapping exercise to develop the draft document using discipline-relevant material both from the Council for the Registration of Forensic Practitioners' forensic podiatry framework and the International Association for Identification's *Forensic Podiatry Role and Scope of Practice* document. It was submitted to the Forensic Science Regulator in June 2013 and circulated within the Forensic Science Regulator's Office for consultation. What followed was a series of amendments and further drafts with much of the content left on the cutting-room floor. A perceived moving of the goal-posts by the Forensic Science Regulator's Office from the group's perspective, and disagreement among members of the group, led to the document's suspension in 2014, coinciding with the appointment of the current Forensic Science Regulator, Dr Gillian Tully.

Meanwhile, publications highlighting the limitations of forensic podiatry expert evidence testimony, particularly in the area of forensic gait analysis, came under the spotlight (Cunliffe and Edmond 2013, 2016, Robinson 2015). Dr Emma Cunliffe and Professor Gary Edmond posed reasonable questions regarding the presentation of gait evidence at trial, however, many of the areas of concern were already being addressed and their evidence base was largely limited to the work of only one forensic gait analyst and focussed on only one case in which evidence was presented by that analyst. Regarding this particular individual, the authors summarised that he 'did not offer clear testimony about the standards that he used to diagnose an abnormal gait feature' (Cunliffe and Edmond 2013, p.344), and highlighted further failings relating to methods used, including a lack of attempts made to limit contextual bias relating to the conclusions drawn.

CREATION OF THE WRITING GROUP FOR FORENSIC GAIT ANALYSIS STANDARDS

Although the Cunliffe and Edmond publications centred on concerns over testimony in court in one trial, by one forensic gait analyst, they spurred a renewed impetus for the completion of the publication of a quality framework document. Reflecting on the difficult previous drafts, the group realised that part of the problem in completion was trying to create a 'one size fits all' set of standards for all areas of forensic podiatry – footprint and feet-in-shoes examinations, podiatry record identification and forensic gait analysis. The group soon came to the conclusion that separate published codes of practice would be more useful. Since the recent criticisms had focussed on the area of forensic gait analysis, this subject area was prioritised over the others in terms of the writing of a standards document. By focussing on only one of the four areas of forensic podiatry, forensic gait analysis had to be considered as a separate entity from forensic podiatry. The release of forensic gait analysis from the original umbrella of forensic podiatry encouraged wider communication and collaboration with the forensic science community. Supported by Dr Anya Hunt, Chief Executive Officer of the Chartered Society of Forensic Sciences, an inaugural meeting of the Forensic Gait Analysis Working Group occurred in May 2017. This offered the opportunity for those working in the field to meet each other, ostensibly to discuss working practice and research. It became clear that although podiatrists formed the majority membership of the new Forensic Gait Analysis Working Group, there were other professions working in the field, such as biomechanists. Research and development discussions aside, the multi-disciplinary members of the forensic gait analysis group were united in that the priority was to develop quality standards for those working in the area of forensic gait analysis. In collaboration with the Chartered Society of Forensic Sciences, the College of Podiatry and the Forensic Science Regulator, a new writing group was formed from members of the Forensic Gait Analysis Working Group to create a quality framework document for forensic gait analysts.

Guided by the Forensic Science Regulator, the new writing group was asked to base the document on the *Codes of Practice and Conduct for Forensic Science Providers and Practitioners in the Criminal Justice System* Issue 3 (Forensic Science Regulator 2016) and International Laboratory Accreditation Cooperation guidance (ILAC G19:08/2014 2014). The writing group also considered forensic science quality assurance in general, how the forensic gait analysis standards document could be developed to reflect credibly the demands of the criminal justice system and how to provide a pragmatic framework for forensic gait analysis providers, particularly as the majority were known to be sole traders. During writing, consideration was also given to the use of the framework and document as a guide to standards in countries other than the UK.

In order to deliver such a framework, an understanding was needed of the principles of quality assurance in the wider forensic community. A quality framework in the context of general forensic science has been succinctly described by Dr Linzi Wilson-Wilde in her article, "The international development of forensic science standards — A review" (2018, p.3).

Dr Wilson-Wilde explains that there are three main areas of the forensic environment in which quality assurance can be applied: (1) the methods employed (2) the forensic practitioner and (3) the facility (e.g. laboratory) (Figure 5.1).

Wilson-Wilde's diagrammatic representation of the relationship between quality assurance and the forensic environment reflects relevant requirements in forensic gait analysis, and the model forms the basis on which forensic gait analysis quality requirements are built. There exists a blurring of boundaries relating to quality assurance terminology. For example, the word 'accreditation' as outlined by United Kingdom Accreditation Service and the Forensic Science Regulator involves the competency of the practitioner in addition to accreditation of the facility, but does not consider professional competency certification on its own. The *Criminal Procedure Rules* Part 19 states that the expert has to show proof of 'accreditation', and defined methods standards are included as part of United Kingdom Accreditation Service accreditation and the Forensic Science Regulator's *Codes*. However, the *Codes* are written in terms of 'Practice and Conduct', implying a set of rules for how professionals should behave, not necessarily about the methods they use. *Criminal Practice Directions* Part 19A (Crim 2018a) discusses methods and how they should be followed but omits the word 'standards'.

In recognition of this somewhat confusing terminology, Wilson-Wilde's representation of the quality assurance framework in forensic science rightly acknowledges that quality assurances of standards, certification and accreditation overlap and therefore presents the model as a Venn diagram (Figure 5.1). In accordance with Wilson-Wilde's model, the following sections, under

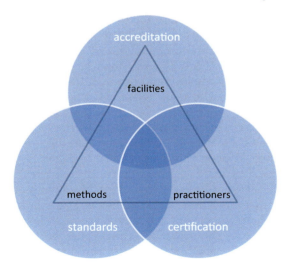

FIGURE 5.1 The relationship between the forensic environment and the quality framework, as described by Brandi and Wilson-Wilde (2013) and further illustrated by Wilson-Wilde (2018, p.3).

the headings 'Methods', 'Practitioners' and 'Facilities', explain how gaps in forensic gait analysis quality assurance have been identified and how strategies have been developed by the writing group as an attempt to plug these gaps.

METHODS

Evidence of various methods used by forensic gait analysis practitioners is apparent in areas of evidence collection and handling, assessments/screening, analyses, comparisons, examinations, interpretations, peer review/critical findings checks and report writing. According to Wilson-Wilde (2018), quality assurance of methods can be assessed through the adherence to standards. Adherence should be demonstrated both at a local level where a facility states its own standard operating procedures, and at a centrally directed level where general standards are adopted. General standards for methods are incorporated in the latest version of the *Codes of Practice and Conduct* developed by the Forensic Science Regulator (2017). Specifically, the *Codes* demand evidence of method validation for specific scientific techniques; in other words, the use of empirically tested research and/or databases. Methods employed should be audited every four years and any new methods should have evidence of validation and be verified by an independent assessor.

The general standards for methods referred to in the Forensic Science Regulator's *Codes* were considered relevant and the group decided they could be mapped across to the forensic gait analysis standards document, whilst commenting in the document on the necessity for adherence to the forensic provider's own standard operating procedures. Debate then ensued among the writing group as to how a sole trader could implement such adherence and the practicalities of a practitioner in charge of its own checks and audit. The term 'forensic unit', frequently used by the Forensic Science Regulator and prevalent in the Forensic Science Regulator's *Codes*, seemed to make sense if the 'unit' referred not only to a whole organisation or a small to medium sized enterprise, but also to a lone practitioner. The inaugural meeting of the Chartered Society of Forensic Sciences' Forensic Gait Analysis Working Group identified that isolation enabled poor practice and that collaboration with other forensic gait analysts would facilitate opportunities for peer review/verification and audit. It was suggested that collaboration and networking in casework could also forge working relationships that would enable preliminary independent assessment of the evidence provided by a commissioning agency (such as the police or Crown Prosecution Service), thus limiting contextual bias. Such relationships would also be beneficial for independent verification of reports.

In November 2017, a month after the draft *Code of practice for forensic gait analysis* document had been submitted to the Office of the Forensic

Science Regulator, the Royal Society published a judicial primer to be utilised by judges presiding over cases in which evidence involving forensic gait analysis is used (2017). The primer was written with the intention of informing the courtroom of the scientific evidence forming the foundations of the methods and techniques employed by the discipline. Although considered by the forensic gait analysis community to be largely useful for a Court of Law, concerns were raised regarding various aspects of the document, including the criticisms of the lack of use of databases published in the clinical literature. Page 16, paragraph 2 of the primer concluded, 'Why such information is rarely, if ever, presented by forensic gait analysts is unclear' (The Royal Society 2017). This conclusion appears to have been predicated on a critique of literature pertaining to clinical gait databases, a key reference of which described a database in which the participants were children. The primer inadvertently answered its own question. Gait data collected from children is of little or no relevance to the vast majority of cases in which forensic gait analysis is used.

Validation of methods used in forensic gait analysis is recognised by the forensic gait analysis community as essential and urgent. Empirical research is currently being carried out to strengthen the underpinning science providing the foundations of forensic gait analysis, including in the area of databases (Van Mastrigt et al. 2018).

PRACTITIONERS

Quality assurance of practitioners can be assessed through education and training, and certification (Wilson-Wilde 2018). Certification of forensic scientists is a recognised method to ascertain the practitioner is maintaining professional standards and assure that they are professionally competent. Practitioners are usually willing to sit assessments for certification purposes, as it is a mandatory requirement for continued employment for many practitioners, especially those in large organisations. This is particularly prevalent in the United States, where certification assessments are provided by many establishments such as the International Association for Identification, the American Board of Criminalistics, the American Board of Forensic Toxicology, etc. For smaller, niche disciplines such as forensic gait analysis, the Chartered Society of Forensic Sciences' Certificate of Professional Competence serves to assess the competency of the practitioner. Anecdotally, there are a very small number of practitioners, as is perhaps the case in other forensic disciplines, who do not share the belief that competency tests are necessary as they consider that their current professional regulation in the healthcare context is sufficient. However, the certificates provide evidence to a court of law that the practitioner is competent to perform an examination and also that they stay abreast of continuing professional development.

FACILITIES

Quality assurance of facilities (sometimes including the competency of employees) can be assessed through accreditation (Wilson-Wilde 2018). The Forensic Science Regulator would have liked to have seen the facilities of forensic science providers fully accredited to the International Standard BS EN ISO/IEC 17025 (or IEC 17020) with scope to include the aforementioned *Codes of Practice and Conduct*, before 2017. Whilst many organisations have completed this requirement, others, particularly medium and small sized forensic provider organisations, have found difficulties, particularly in raising the finances for the accreditation process. Additionally, there is an unknown number of niche forensic providers, many of whom are sole traders. At the time of writing, the cost of accreditation by United Kingdom Accreditation Service for a small laboratory is approximately £12,200 plus yearly charges of £3,303 (UKAS 2019). Clearly, a sole trader acting as an expert witness in one or two cases a year in their niche area of forensic science will not be able to afford such accreditation, or will have to pass the costs of such accreditation onto commissioners, making their services commercially unviable. It is therefore an unachievable goal. The objectives of United Kingdom Accreditation Service accreditation of a laboratory/facility to International Organization for Standardization standards become particularly irrelevant if the sole trader uses a small space in their home as a place to carry out forensic work. This is a typical scenario for forensic gait analysts who require only a PC and screen to carry out their work. Not surprisingly, a survey conducted by the Chartered Society of Forensic Sciences in 2016 determined that of over seventy forensic practitioners, 35% were accredited or working toward accreditation, while approximately 65% were not accredited (McCartney and Amoako 2018). This was despite the Forensic Science Regulator requesting full compliance to United Kingdom Accreditation Service accreditation by 2017.

In recognition of these difficulties, the Forensic Science Regulator has not yet insisted on accreditation for niche disciplines such as forensic gait analysis, archaeology, anthropology, bare or socked footprints and wear features of footwear. Instead, she has requested completion of separate codes of practice for these disciplines, ensuring implementation of an efficient and effective quality management framework for working in the forensic environment. There have been further discussions suggesting that the Chartered Society of Forensic Sciences act as an umbrella organisation for sole traders and small to medium sized enterprises, thus facilitating the previously unachievable United Kingdom Accreditation Service accreditation for niche disciplines. An example of this is the development by the Chartered Society of Forensic Sciences of a generic quality manual to meet the general quality management requirements of the International Organization for Standardization standards (Chartered Society of Forensic Sciences 2019b).

Forensic gait analysis is not the only forensic discipline recognising its own limitations and the need for a code of practice to ensure implementation. For example, in the field of forensic entomology, Gaudry and Dourel note

> Experts are legitimately expected by judges, investigators, lawyers and victims to work in compliance with legal requirements. They must also comply with reliable codes of practice (laboratory and field standards). Hence, the need for the current situation to evolve is a positive way.
>
> **Gaudry and Dourel 2013, p.1031**

DEVELOPMENT OF THE *CODE OF PRACTICE FOR FORENSIC GAIT ANALYSIS* DOCUMENT

At the request of the Forensic Science Regulator, the newly assembled forensic gait analysis multi-disciplinary writing group began the task of developing a forensic gait analysis code of practice document in June 2017. They set themselves an ambitious deadline of submitting the first draft for approval of the Forensic Science Regulator by October 2017, allowing four months for completion. The group was hosted at premises of the Office of the Forensic Science Regulator. The group met three times over the four months to formulate ideas and bring together technical knowledge and understanding of regulatory and legal frameworks, particularly the Forensic Science Regulator's *Codes of Practice and Conduct for Forensic Science Providers and Practitioners in the Criminal Justice System* and International Laboratory Accreditation Cooperation G19. Outside face-to-face meetings, many sections of the draft were circulated electronically for comment and review.

The draft *Code of practice for forensic gait analysis* document was completed on schedule and submitted to the Office of the Forensic Science Regulator in October 2017. It was then circulated for comment in November to the Chartered Society of Forensic Sciences' Forensic Gait Analysis Working Group, the Quality and Competency Manager of the Chartered Society of Forensic Sciences, the Forensic Podiatry Special Advisory Group of the College of Podiatry and the College of Podiatry directorate. The received feedback was useful and varied from comments about the structure of the document to the level of forensic gait analysis evidential value.

During the initial consultation round, members of the Forensic Podiatry Special Advisory Group of the College of Podiatry queried whether or not an additional quality assurance framework was indeed necessary as the Health and Care Professions Council (HCPC) was deemed to be a satisfactory regulatory body for forensic gait practitioners. The query prompted a meeting between the Forensic Science Regulator, Dr Gillian Tully, and the chief executive and registrar of the Health and Care Professions Council, Marc Seale, to discuss the issue. Together, they determined that forensic gait analysts who are registered with the Health and Care Professions Council must comply with their Standards of Conduct, Performance and Ethics, and additionally 'act in accordance with any relevant code of practice or conduct for expert witnesses that sets appropriate requirements in respect of such matters as objectivity, the avoidance of cognitive bias and scientific validity and quality' (Forensic Science Regulator 2018, p.3). It was apparent that HCPC regulations alone are inadequate to assure the quality of a forensic practitioner, as identified by the Chief Executive of the Health and Care Professions Council.

The completed draft was made available for public consultation in July 2018. The consultation period lasted for three months and comments were collated by the Office of the Forensic Science Regulator for further discussion and debate. The majority of responses were received from the office of the Health and Care Professions Council, marking places where it was considered the document lacked detail regarding Health and Care Professions Council regulations. Additionally, feedback pointed to the lack of detail in the document regarding the peer review/verification process in the event two practitioners came to different conclusions. Further discussions ensued relating to EU General Data Protection Regulation and the document was altered again to include GDPR procedures. The *Code of practice for forensic gait analysis* (2019) has now been published, and forms part of the Forensic Science Regulator's *Codes of Practice and Conduct for Forensic Science Providers and Practitioners in the Criminal Justice System* (2017).

The *Code of Practice* is a standalone, self-contained document serving as guidance detailing specific requirements for quality forensic gait analysis provision and is to 'be complied with whenever forensic gait analysis is being undertaken and compliance (or otherwise) declared as part of the expert's report in England and Wales' (Forensic Science Regulator 2018, p.3). In England and Wales, reports written for criminal cases include a set of declarations, according to the *Criminal Procedure Rules* Part 19.4 (j), (k). Declaration 13 states 'I confirm that I have acted in accordance with the code of practice or conduct for experts of my discipline, namely [identify the code]' (Crim 2018b). Forensic gait analysts are now able to insert 'the *Code of practice for forensic gait analysis*' in this section.

The *Code of Practice* document provides guidance for process/methods used, competency and training, validation of methods, responsibilities, cognitive bias, standard operating procedures, document control, peer review, audit, health and safety, technical requirements, reporting and presentation of evidence. It serves as an aid to maintain minimum standards in the forensic arena. Although the main body of the document was completed in just four months, it represents almost thirteen years of discussion, debate, in some areas irreconcilable disagreement and the development of standards, practice and research. Through the Chartered Society of Forensic Sciences' Forensic Gait Analysis Working Group, it is envisaged that further development and understanding will continue, helping the discipline to grow and develop.

REFERENCES

Bellamy, David. 2018. "Quality Manager, CSoFS." Personal communication.

Brandi, J., and L. Wilson-Wilde. 2013. "Standard Methods." In: *Encyclopaedia of Forensic Sciences*, edited by J. A. Siegel, and P. J. Saukko, 522–527. Waltham, MA: Academic Press.

Chartered Society of Forensic Sciences. 2019a. "Certificate of Professional Competence." Accessed July 2019. www.csofs.org/Certificate-of-Professional-Competence-CPC.

Chartered Society of Forensic Sciences. 2019b. "Quality and Competency." Accessed July 2019. www.csofs.org/Quality-Competency.

Chartered Society of Forensic Sciences and College of Podiatry in association with the Forensic Science Regulator. 2019. *Code of practice for forensic gait analysis*, Issue 1. Birmingham: The Forensic Science Regulator.

Cole, S. A. 2007. "Toward Evidence-Based Evidence: Supporting Forensic Knowledge Claims in the Post-Daubert Era." *Tulsa Law Review* 43:263–283.

Crim, P. D. 2018a. *Criminal Practice Directions Division V Evidence. Courts and Tribunals Judiciary*. Strand, London: Royal Courts of Justice.

Crim, P. R. 2018b. "Criminal Procedure Rules Part 19: Expert Evidence. (2015, amended 2018). Courts and Tribunals Judiciary." Strand, London: Royal Courts of Justice.

Cunliffe, Emma, and Gary Edmond. 2013. "Gaitkeeping in Canada: Mis-Steps in Assessing the Reliability of Expert Testimony." *Canadian Bar Review* 92:327–368.

DiMaggio, J. A., and W. Vernon. 2017. *Forensic Podiatry: Principles and Methods*, 2nd ed. Boca Raton, FL: CRC Press/Taylor & Francis Group.

Edmond, G., and E. Cunliffe. 2016. "Cinderella Story? The Social Production of a Forensic "Science"." *Journal of Criminal Law and Criminology* 106:219–274.

Forensic Science Regulator. 2011. "Codes of Practice and Conduct for Forensic Science Providers and Practitioners in the Criminal Justice System, Version 1.0." Home Office. Crown Copyright: 48. Accessed July 12, 2012. https://assets.publishing.service.gov.uk/government/uploads/system/uploads/attachment_data/file/118949/codes-practice-conduct.pdf.

Forensic Science Regulator. 2016. "Codes of Practice and Conduct for Forensic Science Providers and Practitioners in the Criminal Justice System, Issue 3.0." Home Office. Crown Copyright. Accessed September 3, 2017. https://assets.publishing.service.gov.uk/government/uploads/system/uploads/attachment_data/file/499850/2016_2_11_-_The_Codes_of_Practice_and_Conduct_-_Issue_3.pdf.

Forensic Science Regulator. 2017. "Codes of Practice and Conduct for forensic science providers and practitioners in the Criminal Justice System, Issue 4.0." Home Office. Crown Copyright. Accessed April 24, 2018.

https://assets.publishing.service.gov.uk/government/uploads/system/uploads/attachment_data/file/651966/100_-_2017_10_09_-_The_Codes_of_Practice_and_Conduct_-_Issue_4_final_web_web_pdf__2_.pdf.

Forensic Science Regulator. 2018. *Draft Forensic Gait Analysis Code of Practice.* Accessed December 12, 2018. https://assets.publishing.service.gov.uk/government/uploads/system/uploads/attachment_data/file/720883/2018_Forensic_Gait_Analysis_Consultation_Draft.pdf.

Gaudry, E., and L. Dourel. 2013. "Forensic Entomology: Implementing Quality Assurance for Expertise Work." *International Journal of Legal Medicine* 127(5):1031.

Gwinnett, C., J. Cassella, and M. Allen. 2011. "The Trials and Tribulations of Designing and Utilising MCQs in HE and for Assessing Forensic Practitioner Competency." *New Directions in the Teaching of Physical Sciences* 7(7):72–78.

House of Commons Science and Technology Committee. 2005. "Seventh Report of Session 2004-5. Forensic Science on Trial." Science and Technology Committee. London: The Stationery Office Limited.

ILAC G19:08/2014. 2014. "Modules in a Forensic Science Process." Accessed May 14, 2016. www.ilac.org/news/ilacg19082014-published/.

McCartney, C., and E. N. Amoako. 2018. "The UK Forensic Science Regulator: A Model for Forensic Science Regulation?" *Georgia State University Law Review* 34(4):945.

National Research Council. 2009. *Strengthening Forensic Science in the United States: A Path Forward.* Washington, DC: National Academies Press.

Pretty, I. 2006. "The Barriers to Achieving an Evidence Base for Bitemark Analysis." *Forensic Science International* 159(1) (Supplement 1):110–120. doi: 10.1016/j.forsciint.2006.02.033.

Rennison, Andrew. 2012. "Forensic Science Regulator." Personal communication.

Robinson, N. J. 2015. "Forensic Pseudoscience: The Unheralded Crisis of Criminal Justice." Accessed March 16, 2016. http://bostonreview.net/books-ideas/nathan-robinson-forensic-pseudoscience-criminal-justice.

Saks, M. J., and D. L. Faigman. 2008. "Failed Forensics: How Forensic Science Lost Its Way and How It Might Yet Find It." *Annual Review of Law and Social Science* 4(1):149–171. doi: 10.1146/annurev.lawsocsci.4.110707.172303.

Saks, M. J., and J. J. Koehler. 2008. "The Individualization Fallacy in Forensic Science Evidence." *Vanderbilt Law Review* 61(1):199–219.

Skills for Justice. 2014. Accessed January 2014. www.sfjuk.com/.

The Law Commission. 2009. "Consultation Paper No 190. The Admissibility of Expert Evidence in Criminal Proceedings in England and Wales: A New Approach to the Determination of Evidentiary Reliability: A Consultation Paper." Accessed December 18, 2018. http://www.lawcom.gov.uk/app/uploads/2015/03/cp190_Expert_Evidence_Consultation.pdf.

The Law Commission. 2011. *Expert Evidence in Criminal Proceedings in England and Wales.* Vol. 829. Norwich: The Stationery Office.

The Royal Society. 2017. "Forensic Gait Analysis: A Primer for Courts." Accessed November 30, 2017. https://royalsociety.org/~/media/about-us/programmes/science-and-law/royal-society-forensic-gait-analysis-primer-for-courts.pdf.

UKAS. 2019. "Testing Laboratory Small Scope." United Kingdom Accreditation Service. Accessed February 4, 2019. https://www.ukas.com/services/accreditation-services/apply-for-accreditation/what-are-the-costs-of-accreditation/customer-type-testing-laboratory-small-scope/.

Van Mastrigt, N. M., K. Celie, A. L. Mieremet, A. C. C. Ruifrok, and Z. Geradts. 2018. "Critical Review of the Use and Scientific Basis of Forensic Gait Analysis." *Forensic Sciences Research* 3(3):183–193. doi: 10.1080/20961790.2018.1503579.

Vernon, W. 2006. "The Development and Practice of Forensic Podiatry." *Journal of Clinical Forensic Medicine* 13(6):284–287. doi: 10.1016/j.jcfm.2006.06.012.

Vernon, W., J. Walker, S. Reel, H. Kelly, B. Brodie, J. DiMaggio, M. Nirenberg, and N. Gunn. 2010. "The Role and Scope of Practice of Forensic Podiatry." *Journal of Foot and Ankle Research* 3 (S1):O26. doi: 10.1186/1757-1146-3-S1-O26.

Wilson-Wilde, L. 2018. "The International Development of Forensic Science Standards — A Review." *Forensic Science International* 288:1–9. doi: 10.1016/j.forsciint.2018.04.009.

6

Initial contact, preliminary assessment of footage and defining the task

Ivan Birch

COMMISSIONING FORENSIC GAIT ANALYSIS

If you are commissioning forensic gait analysis work for the first time, your first task is to select an appropriate practitioner to do the work. In the UK the National Crime Agency has a database of experts in forensic gait analysis, together with feedback from previous commissioners of work, which should help you decide who to use. The Chartered Society of Forensic Sciences has a register of practitioners who have successfully completed the competency testing process in forensic gait analysis. It might also be worth speaking to other agencies or teams that have used forensic gait analysis to get an insight into their experiences. If you are in any doubt as to the suitability of a practitioner, ask for a copy of their curriculum vitae and a list of recent cases, including the contact details of commissioners of the work, so you can gather some background information.

Once you have selected your analyst, the next challenge is selecting what footage to send them. Sending all the footage you have that shows the figure of interest is not always the best option if much of the footage is unsuitable for use in forensic gait analysis, as it will increase the amount of time taken to do the work, and therefore the cost. If you have not been involved in forensic gait analysis before, knowing what is required in terms of the footage being selected and sent for analysis can be difficult. The following are general criteria for the collection or selection of footage suitable for use in forensic gait analysis. Not all the footage will meet all the criteria listed, but every effort should be made wherever possible.

The footage should show the subject:

+ walking with a natural uninhibited movement (i.e. not affected by other people around them, not carrying objects and not wearing handcuffs)
+ undertaking as many consecutive steps as possible in the same direction
+ from as many angles as possible
+ wearing usual footwear

Please ensure that:

+ the footage will play as sent on any PC or laptop
+ where the footage relies on proprietary software, clear instructions are submitted in conjunction with the footage
+ the footage submitted is limited to the relevant sections of the recordings
+ the footage is of as high a frame rate as possible
+ the footage is accompanied by short explanatory notes confirming the subject of interest and the timings at which they appear in the footage

Where the reference footage does not already exist, please attempt to ensure that:

+ the camera-to-subject angle is matched as closely as possible to that of the questioned footage
+ the subject's walking speed is matched as closely as possible to that of the questioned footage
+ the frame rate is at least 8 frames per second or higher
+ there is a clear line of sight between the camera and subject throughout

- the area where the image is to be captured is well-lit, shadow free and without reflection interference
- the image is not grainy and has good contrast
- the whole of the body is in shot for as many steps as possible
- the subject is wearing clothing that allows a clear view of body and limb movement
- the subject is wearing their own shoes

PROCESS FOR THE ANALYST

Although observational gait analysis has been used for some considerable time in clinical practice, its use in the forensic context is a much more recent development. The research base for its use in the forensic context is growing steadily, and in terms of developing our understanding of process and outcomes, that research base is supplemented by the wealth of gait research that has been undertaken in the clinical context. However, there are significant differences between the use of observational gait analysis undertaken in the clinical context, and observational gait analysis undertaken in the forensic context. In the clinical context observational gait analysis can be adapted and modified to suit the particular needs of each patient, the patient being the important factor, and the gait analysis being a contributor to the wider process of diagnosis and intervention. Although it could be argued that this contribution is analogous to the role of gait analysis in the wider process of identification, in the clinical context the outcome is the product of the overall process and it is the effectiveness of the overall process that is important. In the forensic context each component part is subject to an equal level of scrutiny and is held to an equal level of accountability. The clinical process of diagnosis is convergent, the different strategies involved being used to gain information, possibly with different levels of effectiveness and reliability, leading to a single outcome, established by the clinician. The forensic process is not convergent but parallel. The individual components, forensic gait analysis being one component, all of which have to have been tried and tested, yield independent outcomes, established by the various forensic scientists. The overall outcome is expressly not the responsibility of the forensic scientist, but that of the criminal justice system, and ultimately the court. However, despite this there are clearly lessons that can be learnt regarding observational gait analysis from clinical practice, the most important of which is that the reliability of observational gait analysis is improved if a systematic approach is taken to both the observation and recording of features of gait (Lord, Halligan, and Wade 1998, Rancho Los Amigos National Rehabilitation Center 2001, Perry 2002, Read et al. 2003, Toro, Nester, and Farren 2007a,b, Rathinam et al. 2014, Gor-García-Fogeda et al. 2016). In the forensic context this fact is even more important as it is a fundamental element in the reduction of cognitive bias (see Chapter 11). In forensic gait analysis, information which could inadvertently introduce cognitive bias can come from reading background case information or from seeing both questioned and reference footage before undertaking the analysis of features of gait. While cognitive bias cannot be altogether eradicated from the forensic gait analysis process, its effect can be minimised by taking some simple precautions, one of which is to ensure that the process used for analysing and recording features of gait is the same on every occasion.

The overarching strategy that should be used for forensic gait analysis is referred to as ACE-V (analysis, comparison, evaluation, verification). Triplett and Cooney (Triplett and Cooney 2006) provide a good summary of the development of ACE-V and its application to forensic fingerprint practice. Roy Huber (Huber 1959–1960) first discussed the basic strategy in 1959, drawing the expert witness community's attention to the scientific principle of hypothesis testing as a way of undertaking research and drawing conclusions. Triplett and Cooney noted that the purpose of using this principle was to demonstrate the justification for the support or rejection of a particular hypothesis. The ACE-V strategy has been used by various fields of forensic work since 1959, more recently being adopted for use in fingerprint work, particularly since the advent of Daubert hearings on fingerprint evidence in 2008 (Triplett and Cooney 2006). What ACE-V offers is a sound overarching scientific principle for the organisation of forensic work. Whether or not ACE-V is actually a methodology is debatable, but nevertheless it is a strategy that should always be followed in forensic gait analysis, providing as it does a standardised approach to casework widely utilised and accepted in forensic science. In forensic gait analysis we could say that we use AACE-V (assessment of the footage, analysis of gait, comparison of gait, evaluation of the evidence, verification of the findings). I am not suggesting that we start using this as a new acronym, but we do clearly use a five stage process rather than the four stages denoted by ACE-V.

The following sections of this chapter will lead you through a systematic approach to casework. In general terms this is modelled on the guidance given by the *Code of practice for forensic gait analysis* (2019), which forms part of the quality assurance framework of the UK Forensic Science Regulator (2017).

INITIAL CONTACT

A fundamental element of forensic work is organisation, and your initial contact with a potential commissioner of your forensic services is where it needs to begin. Be sure that you note the contact, the name of the enquirer, the organisation, the date, the method of contact, the task being requested and the outcome of the contact (footage to be sent, no further action etc.). First enquiry to the receipt of footage for preliminary assessment can take days, weeks, months or even years, so you need to be able to backtrack through your records to identify the case, what information was provided by both you and the potential commissioner, and what the requested task was. Emails should be easy to file as long as you have a logical system. Phone contacts are more problematic, but just as important. Keep a secure and legible record of these in whatever medium works for you, remembering that at trial you may be asked for all of this information. Personally, I keep a paper-based list for phone contacts and a separate folder for email contacts. One problem you will find is that if after the initial exchange of information the potential commissioner decides not to progress with the use of forensic gait analysis, they are unlikely to tell you. All information and communication relating to a case, including your initial exchanges with the potential commissioner, must always be stored securely. Electronic documents and data should be protected by effective security, and hard copies should be stored in a locked container.

Your initial contact is most likely to be by phone or email, and in either case it is important to have a standard response ready and waiting. Having a standard response email has a number of advantages. You can take time to pre-prepare the email and therefore ensure that it contains all the necessary and correct information. You can also ensure that there are no typographical errors. You may need to modify the email to suit particular cases, but it will provide a reliable starting point. For telephone contacts you can use the text of the email as the basis of a script, which ensures that you provide all the necessary information.

One thing you will have to be aware and beware of is that irrespective of whether you are being contacted by the prosecution or the defence, they are likely to want to give you a good deal of background information that is not only of no relevance to your work, but can be positively detrimental to you fulfilling your role as an objective expert witness by increasing the potential for cognitive bias. If the enquirer has already included this in their initial email there is little you can do, but if they contact you by phone it is worth stopping them from giving you that background information. What you do need to know from the early exchange of communication is what footage is available, whether or not it shows mid gait steps[1] and what is the task required (i.e. exactly what question is being asked).

The type of task requested or the question being asked can determine the type of report that may result from the enquiry. The type of report required may change as the investigation develops or as a result of the assessment of the footage. Table 6.1 shows four examples of the type of question that can be asked at initial enquiry, the related task and the type of report that might be the final outcome.

If on the basis of this basic information it sounds as if the footage may be suitable for use in forensic gait analysis, you will then need to arrange to receive the footage. This sounds like a simple process, but can be fraught with difficulty. Different commissioners will have different constraints and processes for the conveyance of evidence. Some will insist on delivering the evidence by hand, some will send it by recorded delivery or courier, others will simply post it. In some instances, the footage will be sent by email or

TABLE 6.1
Examples of the Type of Question that can be asked at Initial Enquiry

Question Asked	Task	Report Type
Example 1. We do not have any reference footage. Is the questioned footage suitable for use in forensic gait analysis?	Produce a report detailing your assessment	Screening Report[2]
Example 2. Are the clips showing offences A and B suitable for use in forensic gait analysis? On the basis of their gait, could the figure seen in offence A be the figure seen in offence B?	Produce a report detailing your analysis and conclusion	Investigative Report[2]
Example 3. Is the questioned footage suitable for use in forensic gait analysis? What information can be gained about the gait of the figure in the questioned footage?	Produce a report detailing the findings of your analysis	Investigative Report[2]
Example 4. Is the questioned and reference footage suitable for use in forensic gait analysis? On the basis of their gait, is the figure in the questioned footage the subject in the reference footage?	Produce a report detailing the findings of your comparison	Evaluative Report[2]

third party data transfer platforms. However it gets delivered, there are some basic considerations. Always ask for direct copies of the original footage, and avoiding using the original evidence unless absolutely necessary, which is rarely the case. Specifically ask whether the footage is in its native format, and when you receive it look for indicators that it may have been converted into another format. A direct digital copy of a piece of footage will maintain its quality. Any conversion, however well-meaning, could have resulted in a loss of some elements of the quality, for example frame rate or resolution. I say well-meaning because it is not unusual for several pieces of footage from different sources to have been compiled into a single file to ease playability. Although certainly easier to use, the compilation may not have the same quality as the original footage, which may impact on your ability to use the footage effectively. Incidentally, also bear this in mind at trial. It is very common for all of the footage relating to a case to have been converted into a single compilation to make it easier to play at the trial. The problem here is that at trial the court are no longer looking at the original evidence or possibly the evidence that you used to arrive at your expert conclusions. So during your early communication ensure that you describe exactly what you will need to proceed.

Assuming that you are sent the footage, the next stage is the preliminary assessment of the suitability of the questioned and reference footage for use in forensic gait analysis.

Best practice is for the person undertaking the initial preliminary assessment of the footage not to be the person who is going to be the analyst, and not to be the person who is going to be the verifier. This reduces the potential of cognitive bias, and the potential for decisions regarding the suitability of footage being influenced by the prospect of income generation. While this is certainly best practice, it is not always possible. If you are undertaking the preliminary assessment of the footage and subsequently the analysis, you will need to make this clear to the potential commissioner. You will also need to limit the footage you are sent. To undertake a preliminary assessment of both the questioned and the reference footage would introduce the potential of cognitive bias given that you will have then seen the gait of both the perpetrator and the suspect before you have undertaken any analysis of the questioned footage, something which should not happen in forensic practice. If the use of an independent preliminary assessor is not possible, there is a solution. If you are undertaking both the preliminary assessment and any subsequent analysis, you will have to review only the questioned footage at the outset, best practice being to ensure that the potential commissioner only sends you the questioned footage. If the case moves on to a full analysis, you analyse the questioned footage and only then request the reference footage. This is a perfectly sound strategy from a forensic perspective, but it has one fundamental weakness. You may have invested many hours in analysing the questioned footage and making your notes, only then to find that the reference footage is not suitable for use. Although reference footage is often more easily obtained than questioned footage, it may simply not be possible to obtain any additional footage. Therefore, if you are going to use this strategy you need to explain clearly to the potential commissioner that this is what you are going to do, and that there is the risk that the work may not be able to be completed without additional reference footage.

PRELIMINARY ASSESSMENT OF FOOTAGE

If you are going to undertake forensic gait analysis casework, you will need to give serious consideration to the computer equipment you are going to use. This equipment will play a significant role in your life if you do casework regularly, and will have huge impact on your working conditions and experience. When selecting a computer take professional advice on the choice of processor, bearing in mind that its primary role will be to play video footage. You will need a massive amount of secure storage memory to maintain a complete record of your cases, and most importantly you will need to ensure that all of your work is regularly, safely and securely backed up, preferably on an external hard drive and in more than one place. Monitor size is a personal preference, but high definition large monitors will certainly help you to make the most of the footage you have. Multiple monitors will facilitate the forensic process, allowing you to be playing footage at the same time as making notes on the computer, and accessing necessary reference materials. You will find that a good deal of your time doing casework is actually spent trying to get the footage to play. On some occasions the footage will simply not play, on others the footage will only play at an incorrect speed, or in the wrong aspect ratio, or continuously at normal speed with no option to pause the footage or even select which section you wish to view. This sounds ridiculous, but that is how it is. The software you use to play the footage will in some cases be dictated by the footage itself. Some footage will only play on the proprietary software it was intended for, the software sometimes offering very limited control over the playback. In these cases you are reliant on the commissioner providing you with the appropriate software along with the footage. Even with footage in relatively standard file formats, you will often find that you need to try playing it on a number of pieces of software before it plays satisfactorily, and you will end up with a whole range of video playing software packages installed on your computer. There are software packages available that are specifically intended for playing footage for use in biomechanical assessment, and these often offer a greater range of playback facilities than standard video software, but are of course accordingly expensive. There is also a range of freeware available for video playback, some of which is perfectly adequate for use in forensic gait analysis. Some of the more popular freeware has the advantage that you often find it being used to play the footage when you get to court. Whatever software you choose to use, you will need it to play the footage easily at normal speed, at a variety of slow and fast speeds, and most importantly it must allow you to play the footage frame by frame, preferably forwards and backwards. It is in the provision of these playback facilities that the commercially available biomechanics software has the advantage, being specifically designed for such variations in playback to be a standard mode of use.

Like all forensic processes, the method you use to assess the quality of footage, both in terms of content and technical characteristics, needs to be systematic and repeatable. Birch et al. (2013) developed and tested a simple tool for assessing the quality of footage intended for use in forensic gait analysis (Figure 6.1). The tool is computer-based and automatically yields an overall assessment of the suitability of the footage. The tool provides a basic systematic approach to the assessment, not using technical language, and is therefore suitable for use by anyone involved in a case. The important aspect of the tool is that it alerts whoever is using it to the sort of factors that they need to consider in relation to the use of footage for forensic gait analysis.

The factors that need to be considered can be divided into two groups. The first are the technical qualities of the footage such as resolution, lighting and frame rate. The second are the qualities of the footage related to the figure being observed. The term 'figure' is used here as a general term for the person in the footage. When it comes to casework, I have found it useful to stick to a hard protocol of using the term 'figure' to refer to the person of interest in the questioned footage, and 'subject' to refer to the person of interest in the reference footage. It is worth noting at this point that the use of the word 'individual' should be avoided in casework, the term in the context of forensic science having a very specific meaning associated with the concept that a piece of evidence was produced by a particular source, to the exclusion of all other possible sources (Koehler and Saks 2009). The differential use of figure and subject has developed over a period of time and is based on the experience of the team the author works with. It helps to depersonalise the report and in

CCTV Footage Assessment Tool

Name of User:	
Date:	
Footage Reference:	

| Footage Rating: | |

Ratings are from A (most suitable for use) to E (least suitable for use).

Picture

very sharp	○	○	○	○	○	very blurred
very good contrast	○	○	○	○	○	very poor contrast
very bright	○	○	○	○	○	very dark

Lighting

very good lighting	○	○	○	○	○	very bad lighting
no shadow interference	○	○	○	○	○	significant shadow interference
no reflection interference	○	○	○	○	○	significant reflection interference
direction of light source good	○	○	○	○	○	direction of light source poor

Direction

directly from the side	○	○	○	○	○	directly from the front or back
from below the subject	○	○	○	○	○	from above the subject

Frame Rate

continuous flow of image	○	○	○	○	○	series of still images

Subject

appropriate distance from camera	○	○	○	○	○	far from camera
whole of upper body in shot	○	○	○	○	○	none of upper body in shot
whole of lower body in shot	○	○	○	○	○	none of the lower body in shot
moving very fast	○	○	○	○	○	moving very slowly
10 steps or more in shot	○	○	○	○	○	2 steps or less in shot
clothing good for gait analysis	○	○	○	○	○	clothing poor for gait analysis

Are any of these characteristics so poor as to render the footage unusable in your opinion? ☐

FIGURE 6.1 Tool for assessing the suitability of CCTV footage for use in forensic gait analysis (Birch et al. 2013).

so doing reduce any emotive interpretation of the findings. It avoids confusion and makes it clear in both your notes and in your report that you are not inadvertently assuming that they are the same person. It helps to clarify the report by reducing any confusion about which footage is being referred to, and therefore also helps during report writing and when giving evidence in court. It also removes any possibility of inadvertently using inappropriate terminology in both the report and in the witness box. I have seen expert evidence given, both in report and in testimony, in which the expert referred to the person in both the questioned and the reference footage as the suspect, thereby implying that the perpetrator was the accused, despite the fact that the report was commissioned by the defence. This second group of factors includes camera angle, distance of the figure from the camera, the number of mid gait steps seen and the potential impact of clothing on the observation of features of gait. There may be other case specific factors that need to be considered, but this basic list will provide a basis for assessment. It may also be the case that despite all the other factors being acceptable or even good, there may be a single factor that precludes the use of the footage for what could be described as meaningful gait analysis. For example, the footage may have excellent technical qualities, with the camera angle being good in terms of the perspective of the figure shown, and the figure occupying the majority of the field of view, but not showing any mid gait steps. In this case, although the footage might look good, it is not fit for purpose.

The frame rate of the footage is an important and interesting aspect of the preliminary assessment of footage. The research suggests that the higher the frame rate, the more information analysts can gain from footage regarding gait (Birch et al. 2014). To understand why this is, we need to consider what video footage actually is. Video footage, like film footage, is in fact a series of still pictures. When it is played the still pictures are shown in rapid succession. The brain perceives movement based on this series of still images as a result of the combination of what are referred to as persistence of vision and the phi phenomenon (Birch et al. 2014). Persistence of vision is the continued perception by the brain of a visual stimulus for a short time after the stimulus has ceased (Francis and Grossberg 1996, Johnson, Nozawa, and Bourassa 1998, Hidaka et al. 2010), and the phi phenomenon is the perception of movement based on a series of still images or objects (Steinman, Pizlo, and Pizlo 2000). The perception of movement from CCTV footage is therefore illusionary. Key to how well a piece of footage fools you into believing that it is showing moving images is the frame rate. Footage, video or film, can be captured at any frame rate, but it was established early in the history of moving images that a capture frame rate of around 24 or 25 frames per second will fool the majority of people, and these became the standard frame rates for television, film and video. With advances in technology have come much higher frame rates. As the capture frame rate is reduced, the image eventually begins to be perceived as flickering, the exact point at which this happens varying from observer to observer.

When you watch a piece of footage, your brain makes a series of assumptions as to the way in which an object in the footage gets from one location to another as shown in consecutive frames, filling in the gaps. The degree to which the brain makes these assumptions is a function of frame rate; the lower the frame rate, the less information is actually available to inform the illusion of the moving image in the brain, and the greater the potential for misconstruction.

If footage is played back at the frame rate at which it was captured, the image is perceived as moving at normal speed. If the footage is played back at a frame rate lower than the capture frame rate then the image is perceived as moving in slow motion. It is worth making clear at this point that a piece of footage captured at a low frame rate and then played back at the same frame rate is not time lapse footage, a misapprehension seen in some publications and expert reports the exact origin of which is not known to me. Time lapse footage is footage that has been intentionally captured at a lower frame rate than that at which it is played back, the effect being to make the captured images seem to move much more quickly than they were actually doing so, such as a flower opening in a few seconds. Easily accessible definitions of the technique can be found in the *Oxford Dictionary of English* and at Wikipedia (2019). The importance of frame rate to gait analysis is simple; the more frames that have been captured per step, the more information we have about that step. So if a person is walking slowly at a cadence of 60 steps per minute, and footage of that person is captured at 24 frames per second, we will have 24 frames of information about each step (60 steps per minute/60 seconds in a minute [= 1 step per second] × the number of frames per second). If the capture frame rate is reduced to 8 frames per second, then we would only have 8 frames of information per step to work with. This would reduce the temporal resolution of the footage (the shortest duration of an event at which you could guarantee that you would see the event in the footage). At low frame rates we might fail to capture short duration events, the event occurring entirely between consecutive frames. For example if we captured footage of our person walking at a cadence of 60 steps per minute at one frame per second, when we played the footage back we would see the same point in the gait cycle in every frame, but never see the rest of the gait cycle. Experience suggests that very low frame rates in video footage are becoming rarer, but in the past footage has been submitted for use in forensic gait analysis captured at less than 1 frame per second, one colleague reporting the submission of footage captured at 1 frame per 25 seconds. No minimum frame rate for forensic gait analysis has been proposed, but when considering the suitability of the frame rate for particular gait analysis purposes, it is worth considering the principle of the Nyquist critical frequency, which is the established basis for the determination of the sampling frequency for human movement analysis in the laboratory setting (Durkin and Callaghan 2005). This suggests that the original signal must be sampled at a rate greater than twice the highest frequency of the event of interest. An example in terms of features of gait would be if the heel raise of a figure occurred in two parts, an initial rapid movement lasting for 0.1 of a second, followed by a slower longer movement, the frame rate would have to be 20 frames per second (0.05 seconds between frames) or greater to ensure that the initial rapid movement was seen in the footage. If the frame rate was less than 20 frames per second, the entire initial rapid movement could occur between frames, and not be seen at all in the footage. One suggestion that has been made for dealing with low frame rate footage is the building of a detailed reconstruction of a single step from information collected from a number of steps captured in the footage. The idea is that although all events of the gait cycle will not be seen during one step in low frame rate footage, they may be seen during other steps in the footage. Although this would seem logical from a lay perspective, it is not a biomechanically sound approach, as every step is to varying degrees different, and therefore the context in which the event took place will vary from step to step.

What also needs to be considered is the relationship between the amount of information regarding gait shown by a piece of footage and the cadence at which the person in the footage is walking or running. In essence, the higher the cadence, the fewer frames of information we have for each step. At various walking speeds this variation might be relatively small, but if you are analysing the gait of a person who is running, then you need to consider carefully your estimation of the technical qualities of the footage. If you consider a frame rate of 12 frames per second to be good for analysing walking gait, but you are now analysing a running gait of double the cadence and therefore you have half the effective frame rate, you may need to explain this in both your notes and your report.

In most cases, making a judgement as to the suitability of the footage for use in forensic gait analysis will be relatively straightforward. However, in some cases the footage will be more challenging, having a combination of borderline qualities. Such footage may be suitable for use as corroborative footage, footage that can be used but only in combination with other better footage, but may not be suitable as the sole footage. If in doubt a second informed opinion should be sought, and if still in doubt always err on the side of caution.

Whatever methodology you use to assess the quality of footage, be sure that it is a systematic and repeatable methodology, and be sure that you record both the method used and the results. You may be called upon in court to describe and explain exactly how you assessed the quality of the footage and reached a decision.

Once you have reached a conclusion on the quality of the footage, you need to feed back to the potential commissioner. At this stage you should not have engaged in any gait analysis, just an assessment of the technical quality of the footage. Nevertheless, the technical quality will contribute to the determination of the possible probative value[3] that subsequent analysis could yield. For example, a low resolution piece of footage captured at night may have the potential to provide you with a few basic features of gait, but little in the way of detail. Even if subsequent gait analysis based on the questioned and reference footage yields a good match between the limited number of features discernible in the questioned footage, the probative value will still be relatively low, the poor quality of the footage introducing an element of doubt to your observations that needs to be factored into your conclusions. This may be an important piece of information for the commissioner in informing their decision whether or not to proceed with forensic gait analysis. There may be a variety of possible objectives from the commissioner's perspective, not just the provision of evidence that suggests that the figure in the questioned footage is the subject in the reference footage. These could include exclusion of a person from the investigation, or that the gait analysis does not suggest that it is not the same person. It is the commissioner's decision whether or not to proceed, and your role at this stage is to provide the potential commissioner with as much accurate, impartial and honest information as possible. As with any forensic process, do not overestimate what might be achieved from the footage you have seen when feeding back. At some point in the future you may find yourself in court explaining why you overestimated the potential value of the footage.

Whatever your conclusion with regard to the suitability of the submitted footage, this is a useful opportunity to educate the commissioner. If you think the footage is suitable for use in forensic gait analysis, you will need to explain what the next stage of the process will be. If you think the footage is not suitable for use, you will need to explain why you think it is unsuitable and what constitutes suitable footage. In either case, you are educating the commissioner as to how forensic gait analysis works and what qualities of footage make it more or less suitable. The benefits are twofold. A fully informed commissioner who has had a good, albeit short, engagement with you is more likely to use your services again. They are also much more likely to make informed decisions in the future as to the suitability of footage before they send it, thus reducing the amount of time you spend assessing unsuitable footage.

As was the case with the response to initial contact, well-rehearsed feedback will demonstrate appropriate organisation and reliability, and once again the use of a feedback template can help. In some cases if you have rejected the footage, you will also be asked to provide a brief statement for the commissioner's records as to why the footage was unsuitable, which may be submitted as evidence. Having a standardised method of assessing the footage and taking accurate notes will save you a good deal of time in providing such a statement.

Once you have assessed the footage, reached your conclusion and provided appropriate feedback you may be in for a long wait. This is certainly not always the case, with some cases requiring the submission of a report within 24 hours, but you must be prepared not to hear anything for some considerable time. If the commissioner is for the prosecution, there will certainly be discussions as to the likely probative value of the forensic gait analysis evidence and the balance against the cost implications. If the commissioner is for the defence, there may be similar discussions, but then followed by the necessity of having to apply for legal aid. In either case, you are going to have to prepare an itemised quotation for the work. How much you charge per hour will depend on which country you are based in, but there are some general principles on price setting that you should consider. Your hourly rate should reflect that of other forensic practitioners, including those working in different fields, so find out how much other practitioners charge per hour, and set your rate accordingly. Find out what the rates are that can be gained by defence teams from the legal aid system. If you are truly meeting your obligation to the court of being an independent and objective expert, you should be equally available to both prosecution and defence, and if your rate is beyond that allowed by legal aid, you are not meeting that obligation. Of equal importance is not setting your rate too low. You are offering an expert opinion based on a combination of your training, qualifications and experience, all of which are valuable assets, so do not undervalue your skills. Do not change your rates on a case by case basis. It may be considered unethical to charge different commissioners different rates, and at the very least will not be looked on favourably if a commissioner finds out that you charged another commissioner a cheaper rate.

Break the process down in your quotation and give the number of hours you think it will take you to complete each stage of the work. If another practitioner has undertaken the preliminary assessment, you will need to liaise with them regarding the number of hours it is likely to take you to complete the work, based on their viewing of the footage. If you are working as a sole practitioner, you will at least have seen the questioned footage, and will have to base your quotation on this and an indication from the commissioner as to how much reference footage there is to analyse. Remember that you need to build in the cost of having another practitioner verify your work.

Cases vary considerably in terms of the amount of time they take to complete the gait analysis. Occasionally you will get a case with very limited footage that will take you 10 to 20 hours to complete, and occasionally you will get a case with a large number of pieces of footage that will take you 60 to 70 hours or even more. But most cases will tend to fall between 20 and 40 hours including verification. Do not over-quote, but certainly do not under-quote. The number of hours shown above draws attention to the fact that undertaking forensic gait analysis casework is a serious commitment in terms of time and should not be entered into unless you are able to commit to dedicating the appropriate amount of time within the required timescales.

If your quotation is accepted, wait until you receive an official confirmation before starting on the work.

DEFINING THE TASK

Having obtained a useable piece of footage and having been given approval to proceed with the work, the next thing you need to do is agree the task. You will have the benefit of science-based education and will therefore understand the idea of objectivity, theory testing and the need to identify and enunciate in an appropriate way the question being asked. Not everyone involved in a case will have such a background and therefore such specific understanding. This is not a criticism, just a fact. Look carefully at the instruction and the question being asked and ensure that they are appropriate. How could they be inappropriate? We have previously received instructions that, on the list of descriptions of the questioned footage, give the name of the suspect as being seen in each piece of footage. We have also received instructions that ask the expert to prove that the figure in the questioned footage is the subject in the reference footage. In the majority of cases, the question being asked is in essence obvious: is the figure in the questioned footage the subject in the reference footage? Nevertheless, the question may still need some refinement before you embark on the analysis. Be sure that the question asked can be appropriately included in your report, bearing in mind that your responsibility is to give the court an unbiased and objective expert opinion. If necessary, contact the commissioning agency and have a conversation regarding the exact nature of the question being asked and agree the exact wording.

One particular issue relating to the appropriateness of the question being asked that occurs regularly is the relationship to each other of several pieces of questioned footage. The information provided may make the assumption that the, let us say, five pieces of questioned footage are of the same figure, and can therefore be used in conjunction with each other to determine the features of gait that characterise the gait of the figure. As the expert analyst that may ultimately be in the witness box, you need to establish how the commissioner knows that the five pieces of footage show the same figure. In the forensic context, the fact that the figure in all five pieces of questioned footage appears to be wearing the same clothes, and appears to be the same height and build, is of no use to you for a number of reasons. You do not know how many people in that location wear the same clothes, perhaps as a uniform or fashion, and are roughly the same height and build. It is possible that a group of perpetrators could be exchanging clothes to cause confusion. Importantly, unless you have had specific training in clothing identification, body mapping and height estimation, you are not qualified to make such a judgement; it is outside the scope of your expertise. You will therefore need to ascertain how the commissioner of the work determined that it is in fact the same figure in each piece of footage. If they say that they have another expert who will testify that they are the same figure based on the application of their expertise, or that the timings of the footage are continuous or overlap, perfect. If they tell you that a police officer will testify that they are the same figure because they recognise them, it is not perfect. In the latter case, you could find that having done all of the work, produced your report and attended court, your evidence is disallowed because it was based on an unsubstantiated premise, or on an identification that is itself disallowed by the court.

To the commissioner, unfamiliar with how forensic gait analysis works, the difference between five separate pieces of footage and five related pieces of footage may not be obvious, but it is of fundamental importance. Suppose our five pieces of footage each show 10 different mid gait steps. If there is a strong piece of evidence that the five pieces of footage show the same figure, such as the footage showing consecutive and overlapping time periods from different cameras, then the analyst can amalgamate the information gained from the five pieces of footage, and the features of gait identified as being characteristic of the gait of the figure will be based on 50 mid gait steps. The amalgamated information can then be compared with the gait of the subject in the reference footage. If the five pieces of footage cannot be linked with certainty, then features of gait will have to be considered to be characteristic of the gait of the figure in each piece of footage separately, each being based on only 10 mid gait steps. Five comparisons will then have to be made with the gait of the subject in the reference footage. The probative value of the comparisons made with the amalgamated footage is likely to be greater than that of each of five comparisons based on the separate pieces of footage. There will also be a fundamental difference in the structure of your report, which in the latter case will have to include five separate outcomes, which could all be different.

Figure 6.2 gives a summary of the work flow during the preliminary stages of the forensic gait analysis process.

Once the preliminary assessment of the footage has been completed and the task agreed, confirm the requested task in writing with the commissioning agency to ensure both parties fully understand the question being asked

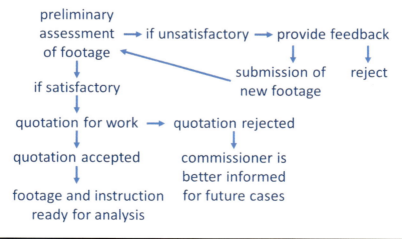

FIGURE 6.2 Summary of the initial stages of the forensic gait analysis process.

and the final product required. Having confirmed the task and the wording of the specific question being asked, you can move on to doing the actual gait analysis.

NOTES

1. A mid gait step is a purposeful step taken at a usual speed during a sequence of consecutive steps taken in approximately the same direction. The important factor about a mid gait step in the forensic context is that it is the type of step that is most likely to represent the usual gait of the person being analysed. During the first few steps taken, the mass of the body has to be accelerated to reach the usual walking speed, and during the last few steps, the mass of the body has to be decelerated. These steps can therefore be regarded as being different in function from those during the middle of a sequence of steps, the mid gait steps. Of course every step is perhaps slightly different in terms of mechanics, but using mid gait steps is most likely to yield a good representation of usual gait. I have said 'purposeful step' to rule out the sort of steps that can sometimes be seen being taken when someone is for example waiting and wanders around in the vicinity in an aimless fashion to kill time, or when someone is agitated. I have said 'taken in approximately the same direction' because any sequence of steps is likely to include subtle perturbations in terms of the line of progression. What needs to be avoided is the use of steps during which there is a sudden change of direction that requires a substantial change in the mechanics of gait. As with accelerative and declarative steps, such steps are less likely to be representative of usual gait.
2. A screening report is a report that records the preliminary assessment of the footage in terms of its suitability for use in forensic gait analysis. An investigative report is a report that provides information that helps the commissioner make progress with an investigation, such as providing investigative leads. An evaluative report is a report that records your comparison of the features of gait and your expert assessment of the probative value of the gait evidence.
3. Probative value is the degree to which a piece of evidence can be relied upon to demonstrate that a disputed point is or is not true.

REFERENCES

Birch, I., W. Vernon, J. Walker, and J. Saxelby. 2013. "The development of a tool for assessing the quality of closed circuit camera footage for use in forensic gait analysis." *Journal of Forensic and Legal Medicine* 20(7):915–917. doi: 10.1016/j.jflm.2013.07.005.

Birch, I., W. Vernon, G. Burrow, and J. Walker. 2014. "The effect of frame rate on the ability of experienced gait analysts to identify characteristics of gait from closed circuit television footage." *Science and Justice* 54(2):159–163. doi: 10.1016/j.scijus.2013.10.002.

Chartered Society of Forensic Sciences and College of Podiatry in association with the Forensic Science Regulator. 2019. *Code of practice for forensic gait analysis*, Issue 1. Birmingham: The Forensic Science Regulator.

Durkin, Jennifer L., and Jack P. Callaghan. 2005. "Effects of minimum sampling rate and signal reconstruction on surface electromyographic signals." *Journal of Electromyography and Kinesiology* 15(5):474–481.

Forensic Science Regulator. 2017. "Codes of practice and conduct: For forensic science providers and practitioners in the criminal justice system, issue 4." Birmingham: The Forensic Science Regulator.

Francis, Gregory, and Stephen Grossberg. 1996. "Cortical dynamics of form and motion integration: Persistence, apparent motion, and illusory contours." *Vision Research* 36(1):149–173.

Gor-García-Fogeda, María Dolores, Roberto Cano de la Cuerda, María Carratalá Tejada, Isabel Mª Alguacil-Diego, and Francisco Molina-Rueda. 2016. "Observational gait assessments in people with neurological disorders: A systematic review." *Archives of Physical Medicine and Rehabilitation* 97(1):131–140. doi: 10.1016/j.apmr.2015.07.018.

Hidaka, Souta, Wataru Teramoto, Jiro Gyoba, and Yôiti Suzuki. 2010. "Sound can prolong the visible persistence of moving visual objects." *Vision Research* 50(20):2093–2099.

Huber, R. 1959–1960. "Expert witness." *The Criminal Law Quarterly* 2:276–295.

Johnson, Brady, Georgie Nozawa, and Charles M. Bourassa. 1998. "Differences in the luminance of the first and second displays affects visible persistence in opposite ways." *Vision Research* 38(9):1233–1238.

Koehler, Jonathan J., and Michael J. Saks. 2009. "Individualization claims in forensic science: Still unwarranted." *Brooklyn Law Review* 75:1187.

Lord, S. E., P. W. Halligan, and D. T. Wade. 1998. "Visual gait analysis: The development of a clinical assessment and scale." *Clinical Rehabilitation* 12(2):107–119.

Perry, Jacquelin. 2002. *Gait Analysis*. Vol. revised edtion (2001), *Normal and Pathological Function*. Downey, CA: Los Amigos Research & Education Center. Original edition, 1992.

Rancho Los Amigos National Rehabilitation Center. 2001. *Observational Gait Analysis Handbook*. Vol. 4th. Downey, CA: Los Amigos Research & Education Institute.

Rathinam, Chandrasekar, Andrew Bateman, Janet Peirson, and Jane Skinner. 2014. "Observational gait assessment tools in paediatrics - A systematic review." *Gait and Posture* 40(2):279–285. doi: 10.1016/j.gaitpost.2014.04.187.

Read, Heather S., M. Elizabeth Hazlewood, Susan J. Hillman, Robin J. Prescott, and James E. Robb. 2003. "Edinburgh visual gait score for use in cerebral palsy." *Journal of Pediatric Orthopaedics* 23(3):296–301.

Soanes, C. 2003. *Oxford Dictionary of English*. Oxford: Oxford University Press.

Steinman, Robert M., Zygmunt Pizlo, and Filip J. Pizlo. 2000. "Phi is not beta, and why Wertheimer's discovery launched the Gestalt revolution." *Vision Research* 40(17):2257–2264.

Toro, Brigitte, Christopher J. Nester, and Pauline C. Farren. 2007a. "The development and validity of the Salford Gait Tool: An observation-based clinical gait assessment tool." *Archives of Physical Medicine and Rehabilitation* 88(3):321–327.

Toro, Brigitte, Christopher J. Nester, and Pauline C. Farren. 2007b. "Inter- and intraobserver repeatability of the Salford Gait Tool: An observation-based clinical gait assessment tool." *Archives of Physical Medicine and Rehabilitation* 88(3):328–332.

Triplett, Michele, and Lauren Cooney. 2006. "The etiology of ACE-V and its proper use: An exploration of the relationship between ACE-V and the scientific method of hypothesis testing." *Journal of Forensic Identification* 56(3):345.

Wikipedia. 2019. "Time-lapse photography." Accessed July 2019. https://en.wikipedia.org/wiki/Time-lapse_photography.

Analysing the questioned and reference footage

Ivan Birch

D espite the number of times the word gait is used in this chapter, the amount of time spent doing gait analysis for a case in forensic practice is often relatively small compared to the overall amount of time invested in the case as a whole.

Before you actually get to the questioned footage, you will need to have identified what methodology you are going to use for observing and recording the features of gait. The guidance given by the UK Forensic Science Regulator's *Codes of Practice and Conduct* (2017), or the equivalent document in your own country of practice, should inform your choice, as should the available peer-reviewed research. Forensic science in general has been criticised for not adopting a scientific approach to practice, and this criticism is certainly warranted in some cases. The method you use should be logical, systematic, repeatable, reproducible, and should have been tested and the results published. Some of the observational gait analysis tools developed in the clinical context certainly meet these criteria (Lord, Halligan, and Wade

1998, Rancho Los Amigos National Rehabilitation Center 2001, Toro, Nester, and Farren 2007a,b). However, these tools are usually intended for use in relation to a specific subject group or pathology, and as such are not directly applicable to forensic use. Nevertheless, they can provide the basis of a useable tool. Birch et al. (2019) described the testing of the Sheffield Features of Gait Tool (see Appendix 1) which was specifically designed and tested for use in forensic gait analysis. As we have already discussed, the research has shown that the reliability of observational gait analysis improves if the method used is systematic (Portney and Watkins 1993, Lord, Halligan, and Wade 1998, Perry 2002, Read et al. 2003, Toro, Nester, and Farren 2007a,b, Rathinam et al. 2014, Gor-García-Fogeda et al. 2016). The method you use should lead you through a whole body gait analysis, which would generally work from the head to the feet. Whatever tool or methodology you use, you should use exactly the same process for each piece of footage in a case.

An important aspect of forensic work is the notes that are taken. All forensic work requires the generation of contemporaneous notes which can be requested at trial. Notes typed directly into your computer are preferable to handwritten notes, and as with many other parts of the forensic process, the generation and subsequent use of a standard template for your notes will save time and ensure consistency. Whatever the method by which notes are made, they should not be altered once recorded. So during all stages of the forensic gait analysis process you need to keep detailed but succinct, legible and accurate notes. Before you do the gait analysis, there are a number of things that need to be recorded in your notes, including:

a) the case information
b) the date and time of the analysis
c) if the footage is stored on a DVD, the exact labelling on the disc
d) if the footage is stored on a data stick, the name (if any) of the data stick
e) the name of the file in which the piece of footage is stored
f) any identifying labelling intrinsic to each piece of footage (e.g. camera number or reference)
g) the camera position, relative to the field of view (e.g. below head height, above head height, well above head height)
h) whether the footage appears to have been captured during daylight or darkness. This can affect both the technical quality and visible content of the footage and can therefore result in the same camera producing two pieces of footage that vary in their suitability for use in forensic gait analysis.
i) a brief description of what can be seen in terms of general background in each piece of footage. Working around the field of view describing what you can see will help you to immerse yourself in the footage. This will help you to identify factors that might affect the gait of the figure/subject, such as slopes or uneven surfaces.

j) an estimation of the resolution and lighting in terms of the suitability of the footage for use in gait analysis. Unless you have training in video analysis this should be couched in non-technical terms, such as excellent, very good, good, adequate, poor. Note that this scale is skewed towards the better end of the scale as the footage you are now considering has already been vetted through preliminary examination, so there should not be any unsuitable footage in terms of resolution or lighting.
k) an estimation of the frame rate at which the footage was captured. The easiest way to get a reasonable estimation of the frame rate is to move the footage to a frame at which the timer moves to a new second, and then click through the footage one frame at a time, counting only the frames that show a change in the position of a moving object or figure, until the counter moves to the next second. It is important only to count the frames that show a change in position of a moving object or figure as in some formats, particularly if the format has been converted, there are always the same number of frames per second in the recording, but not all of them are different, each frame being repeated several times. You are only interested in the number of frames per second that show movement. Although some software players will show the frame rate, it is always preferable to check for yourself. They may be counting the number of frames, rather than the number of frames that show movement.
l) the time at which the figure/subject enters the field of view
m) where the figure/subject enters the field of view
n) the number of steps and the number of mid gait steps seen. This is an important note to make. Commissioners and judges will often describe the footage in terms of its length in time. Of far more importance for gait analysis is how many steps are fully or partially visible. The steps can be divided down into subsets:

partially seen – you will often see part of a step, perhaps mid stance to initial contact, or toe off to mid stance, in footage as the figure/subject enters the field of view or emerges from behind an object or another person. There may also be steps where part of the body cannot be seen, but which can yield valuable information regarding those parts that can be seen. While these partially seen steps are not good for basing absolute conclusions on as you are partially missing the context of the step, they may be useful in corroborating information that you get from the fully seen steps.

fully seen – fully seen steps provide more certain information and will form the basis of your analysis

mid gait steps – mid gait steps are steps taken consecutively at approximately the same speed and approximately in a straight line. These are going to be the backbone of your gait analysis as they are the steps that are most likely to represent accurately the usual gait of the figure/subject.

non-mid gait steps – the steps taken during acceleration, deceleration and changing direction might provide you with some corroborating information, but as noted above are much less likely to represent the usual gait of the figure/subject. Record the number of the various steps chronologically throughout the footage. It is possible that you will get asked for this information when giving evidence at trial as the court tries to understand exactly how much information you had to work from.

o) any factors that might affect the gait of the figure/subject. In some pieces of footage this can be a long list, but when later considering the probative value that can be placed on the various pieces of footage, this information can be critical. Such factors include:

footwear, or lack of it – not often an issue in questioned footage, but sometimes a real issue with reference footage. It is not uncommon for the shoes of subjects in custody to be removed, which can of course affect their gait. When the footwear has been removed you can usually observe this in the footage, as long as you are looking for it. A more complex occurrence is when the subject's footwear has been taken, and they have been given another pair of shoes. Again, this could significantly affect their gait, particularly as the issued footwear is usually slip-on, is not necessarily the correct size, and is often worn with the backs of the shoes folded down.

slope of the ground – walking on a slope, whether up or down or across it, can affect gait. However, there are a number of published papers that have investigated the nature of the effect (Redfern and DiPasquale 1997, Leroux, Fung, and Barbeau 2002, Prentice et al. 2004, Damavandi, Dixon, and Pearsall 2012). The important point is that if you have noticed that the ground is sloping, you have acknowledged it in your notes, and if necessary in your report. If you are not sure about the terrain or slope you can see in a piece of footage, a useful approach can be to identify where the footage has been captured and visit that location on a virtual platform such as Google Maps. If necessary, visit the location in person.

handcuffs – again, obviously not an issue with questioned footage, but sometimes an important consideration in reference footage, as the subject arrives at the custody suite. Arm swing is an important mechanical part of gait, its contralateral exertion of force on the body working in opposition to that imposed by the lower limbs. This loss of arm swing can affect gait (Umberger 2008,

Collins, Adamczyk, and Kuo 2009, Bruijn et al. 2010, Meyns, Bruijn, and Duysens 2013).

proximity of other figures – as with many aspects of gait, the exact nature of the effect that the proximity of other people will have on gait will vary from person to person and from situation to situation. Nevertheless, there is likely to be an effect. If you are walking with someone or following someone, then you are likely to adjust your gait to match their speed, if you are walking in the opposite direction to someone you are likely to alter your gait to avoid a collision, and if you are walking with a group of people, you are likely to do both. If you are also talking to someone while you are walking, it can also affect the way you position and move your head and torso.

carrying objects – the obvious effect here is that caused by carrying a heavy object on the mass and inertia of the body. However, there are other effects related to changes in the movement of the arms if an object is carried in one hand and not the other, or on the position of the torso if the object is carried on the back. An interesting effect here is that caused by carrying an object such as a weapon in both hands in front of the body. Even if the object has no significant mass, the fact that the hands are in effect joined together can affect their ability to counteract the contralateral movement of the lower limbs, thereby affecting gait. An object does not therefore have to be heavy to affect gait.

These are a few examples of factors that can affect the gait of the figure/subject seen in the footage, and there are many others that you may need to identify and acknowledge in your notes and report. In all cases, the exact nature of the effect that a particular factor has on the gait of the figure/subject cannot be predicted or even determined from a piece of CCTV footage. The important point is that you have identified factors that could potentially have affected the gait of the figure/subject and have taken these potential effects into account later in the process when you are comparing the gait of the figure in the questioned footage to that of the subject in the reference footage.

p) the time at which the figure/subject leaves the field of view
q) where the figure/subject leaves the field of view

Considering and recording all of this information will help you to identify the technical and content limitations of the footage in terms of its use in forensic gait analysis.

We have now, at last, arrived at the part where you get to do some actual gait analysis. As has already been emphasised, it is important that the gait analysis is conducted in a systematic and repeatable fashion, and the use of a recognised gait analysis tool will facilitate this approach. Whatever tool you choose to use, work through it in the order it is laid out and do not skip from section to section as you see features of gait that draw your interest. Doing so will jeopardise the objectivity and reliability of your analysis. You will inevitably have to play each piece of footage multiple times, each time concentrating on one aspect or feature of gait. To gain the maximum amount of detailed information about each feature of gait, you will have to play the footage at varying speeds and in some cases frame by frame. Some features of gait, such as the overall symmetry of the gait, are more easily seen when the footage is played at full speed, while others will require a frame by frame approach. Playing footage through frame by frame has the advantage of minimising the potential effect of the brain trying to fill in the gaps between consecutive frames of information. So even if a feature of gait can be seen when the footage is playing at full speed, it is worth also viewing it frame by frame or at least in slow motion to ensure that part of what you are seeing is not an illusion. Complete your analysis of each piece of the questioned footage in turn, recording it in your notes as you do the analysis. Take your time and play the footage as many times as it takes to reach a conclusion regarding each feature of gait. Some features will be obvious, others will not. If in doubt err on the side of caution, and if a feature cannot be determined in a piece of footage then say so. One of the guiding principles based on the experience of the author is 'when I play this footage in court, will the jury be able to see what I can see?' You could of course take the approach that you are the expert, and therefore what you say goes. However, such an approach rarely carries any weight in court, where the role of the expert is to help the trier of fact understand the evidence. Your expertise should be used to guide the court through your evidence, which is of course entirely based on what can be seen in the footage. If you cannot show the court the feature of gait you are talking about in the footage, you are unlikely to convince the court that it can be seen at all. Saying that you can see something regarding gait that you cannot show the court in the footage, no matter how minor, could result in you inadvertently undermining the credibility of all of your evidence, and in some cases the credibility of other pieces of evidence relating to the case.

As you complete the analysis and note-taking for each piece of footage, avoid referring to the notes from previous pieces of footage. This will help minimise the effects of cognitive bias, and will result in a more reliable overall assessment of the gait of the figure. If you are doubtful about your observation of a particular feature of gait in the piece of footage you are dealing with, it is very tempting to go back to a previous piece of footage to corroborate your observation. Do not do this. Treat each piece of footage as a separate piece of evidence, and record exactly what you see in the footage, not what you've seen in the previous pieces of footage. It is very likely that there will be variations in your observations from one piece of footage to another, even when you know from information provided that it is the same person. Steps vary in function (e.g. acceleration, deceleration, turning, correction of errors made on the previous step) and the exact context of individual steps will result in variations in features of gait. You should not expect all features of gait to be the same or to be exhibited on every step. Some features of gait exhibited by a person will be constant, occurring on every step. Others may occur regularly or irregularly, but not necessarily frequently. There is a popular misconception that features that occur on every step are more valuable in the forensic context than those that occur at intervals. This is not necessarily the case. An unusual feature that occurs on a number of occasions in a piece of footage and in a number of pieces of footage, may prove to be of considerable probative value when drawing conclusions from your comparison. Always record exactly what you see in the footage, and if you are not sure what you can see, then record your doubt as well as the observation. Variation is a natural part of human gait, and therefore variation can be as characteristic of a person's gait as any other feature. An erratic line of progression can be a characteristic feature of a person's gait, as can the irregular presentation of a particular variant of a feature of gait. If you see variation, record it. You may later find exactly the same variation in the reference footage.

Having completed the analysis and note-taking for each piece of the questioned footage, you will now have a number of lists of features of gait, one for each piece of footage. You now have to summarise your observations and produce a list of features of gait that the footage as a whole suggests is characteristic of the gait of the figure. The more systematic your approach has been to the analysis of each piece of footage, the easier this stage of the process will be. The Sheffield Features of Gait Tool (Birch et al. 2019) (see Appendix 1) uses a spreadsheet approach, the features of gait and their variants being listed on the left side of the screen, with individual columns to the right of the screen for completion for each piece of footage. Previously completed columns are hidden from view during the completion of subsequent columns, reducing their impact on the recording of features from later pieces of footage. Once all the questioned footage has been analysed, all the completed columns can be revealed, aiding the summary process. The first consideration is in how many pieces of footage has a feature of gait been observed. If you have analysed 10 pieces of questioned footage, and the same variant of a feature of gait has been seen in all 10 pieces of footage, your summary for that piece of footage is straightforward. However, it is rarely that simple. A variant of a feature of gait may have been seen in six of the pieces of footage, and a different variant in the other four. The solution is not necessarily just to summarise with the variant that occurred most often, although this may be the final outcome. You need to consider the value of each of the pieces of footage in terms of its ability to demonstrate the usual gait of the figure. Some pieces of footage will show more mid gait steps than others. Some may show other figures in proximity to the figure in question that may be affecting their gait. The figure in question may be carrying an object or changing direction in the middle of the sequence of steps shown. All these factors and potentially many more need to be considered when estimating the value of each piece of footage, and therefore the weight you can place on the observations you have made from that footage. In the case of our differing observations from the 10 pieces of footage, the four pieces of footage may each show more than 20 consecutive mid gait steps taken in a relatively straight line, while the six each show less than 10 steps taken with an obvious variation in the line of progression. You may therefore consider the variant of the feature of gait seen in the four pieces of footage to be more indicative of the usual gait of the figure/subject than that seen in the six pieces of footage.

A further complexity is added when a feature of gait cannot be observed in a piece or several pieces of footage, perhaps due to the camera angle relative to the figure. Again, the relative value of the pieces of footage where the feature has been seen has to be considered relative to the uncertainty introduced by not being able to make a comparable observation in other pieces of footage.

At the time of writing, research is being undertaken into the generation and use of frequencies of observation based on the observations made from multiple pieces of footage, which can then be used to represent the certainty that the summary observation made is truly representative of the gait of the person being analysed. The work is in the early stages of development, but offers the possibility of the development of a more objective approach to summarising features of gait observed in multiple pieces of footage.

What has not been considered so far is how to record and then summarise features of gait that do not occur on every step. One method is to have a set of notations that can be added to the observation of a feature of gait. The system used by the authors is to record such features as occurring on the majority of steps, on some steps or on the minority of steps. This allows for subtle variances in observations from individual pieces of footage, and in the summary of collective observations, allowing you to record when two variants of a feature of gait are both seen to occur. Such an approach improves the accuracy of observation and recording, and can improve the probative value of subsequent comparisons.

Once you have completed your analysis, note-taking and summarisation from the questioned footage, you can then move onto the reference footage. Unless there is a very good reason not to do so, you should leave a reasonable time period between analysing the questioned footage and analysing the reference footage, to reduce the potential for what you have observed in the questioned footage influencing what you perceive to observe in the reference footage. Starting your analysis of the reference footage at least on the following day from completing the analysis of the questioned footage is a good strategy if possible. In view of the subjective nature of observational gait analysis, and the associated potential for cognitive bias, this is fertile ground for cross-examination questions in court, and I have been asked at trial how long the gap was between the examination of the questioned footage and that of the reference footage.

Use exactly the same process for the reference footage as you used for the questioned footage, multiple plays of each piece of footage, at varying

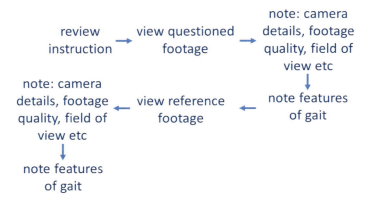

FIGURE 7.1 Summary of the analysis stages of the forensic gait analysis process.

speeds, working systematically through the features of gait for each piece of footage. Importantly you must not refer back to either the questioned footage, or your notes relating to the questioned footage, while you are analysing the reference footage. To do so would introduce cognitive bias.

Figure 7.1 gives a summary of the work flow during the analysis stages of the forensic gait analysis process.

Once the reference footage has been analysed and summarised, the next step is to undertake the comparison of the gait that you concluded represents the usual gait of the figure in the questioned footage and that of the subject in the reference footage.

REFERENCES

Birch, Ivan, Maria Birch, Lucy Rutler, Sarah Brown, Libertad Rodriguez Burgos, Bert Otten, and Mickey Wiedemeijer. 2019. "The repeatability and reproducibility of the Sheffield Features of Gait Tool." *Science and Justice* 59(5):544–551.

Bruijn, Sjoerd M., Onno G. Meijer, Peter J. Beek, and Jaap H. van Dieën. 2010. "The effects of arm swing on human gait stability." *Journal of Experimental Biology* 213(23):3945–3952.

Collins, Steven H., Peter G. Adamczyk, and Arthur D. Kuo. 2009. "Dynamic arm swinging in human walking." *Proceedings of the Royal Society. Series B: Biological Sciences* 276(1673):3679–3688.

Damavandi, Mohsen, Philippe C. Dixon, and David J. Pearsall. 2012. "Ground reaction force adaptations during cross-slope walking and running." *Human Movement Science* 31(1):182–189.

Forensic Science Regulator. 2017. "Codes of practice and conduct: For forensic science providers and practitioners in the criminal justice system, issue 4." Birmingham: The Forensic Science Regulator.

Gor-García-Fogeda, María Dolores, Roberto Cano de la Cuerda, María Carratalá Tejada, Isabel Mª Alguacil-Diego, and Francisco Molina-Rueda. 2016. "Observational gait assessments in people with neurological disorders: A systematic review." *Archives of Physical Medicine and Rehabilitation* 97(1):131–140. doi: 10.1016/j.apmr.2015.07.018.

Leroux, Alain, Joyce Fung, and Hugues Barbeau. 2002. "Postural adaptation to walking on inclined surfaces: I. Normal strategies." *Gait and Posture* 15(1):64–74.

Lord, S. E., P. W. Halligan, and D. T. Wade. 1998. "Visual gait analysis: The development of a clinical assessment and scale." *Clinical Rehabilitation* 12(2):107–119.

Meyns, Pieter, Sjoerd M. Bruijn, and Jacques Duysens. 2013. "The how and why of arm swing during human walking." *Gait and Posture* 38(4):555–562.

Perry, Jacquelin. 2002. *Gait Analysis*. Vol. revised edtion (2001), *Normal and Pathological Function*. Downey, CA: Los Amigos Research & Education Center. Original edition, 1992.

Portney, L. G., and W. P. Watkins. 1993. *Foundations of Clinical Research: Applications to Practice*. Stamford: Lange.

Prentice, Stephen D., Erika N. Hasler, Jennifer J. Groves, and James S. Frank. 2004. "Locomotor adaptations for changes in the slope of the walking surface." *Gait and Posture* 20(3):255–265.

Rancho Los Amigos National Rehabilitation Center. 2001. *Observational Gait Analysis*. Vol. 4th. Downey, CA: Los Amigos Research & Education Institute.

Rathinam, Chandrasekar, Andrew Bateman, Janet Peirson, and Jane Skinner. 2014. "Observational gait assessment tools in paediatrics - A systematic review." *Gait and Posture* 40(2):279–285. doi: 10.1016/j.gaitpost.2014.04.187.

Read, Heather S., M. Elizabeth Hazlewood, Susan J. Hillman, Robin J. Prescott, and James E. Robb. 2003. "Edinburgh visual gait score for use in cerebral palsy." *Journal of Pediatric Orthopaedics* 23(3):296–301.

Redfern, Mark S., and James DiPasquale. 1997. "Biomechanics of descending ramps." *Gait and Posture* 6(2):119–125.

Toro, Brigitte, Christopher J. Nester, and Pauline C. Farren. 2007a. "The development and validity of the Salford Gait Tool: An observation-based clinical gait assessment tool." *Archives of Physical Medicine and Rehabilitation* 88(3):321–327.

Toro, Brigitte, Christopher J. Nester, and Pauline C. Farren. 2007b. "Inter- and intraobserver repeatability of the Salford Gait Tool: An observation-based clinical gait assessment tool." *Archives of Physical Medicine and Rehabilitation* 88(3):328–332.

Umberger, Brian R. 2008. "Effects of suppressing arm swing on kinematics, kinetics, and energetics of human walking." *Journal of Biomechanics* 41(11):2575–2580.

8

Comparison of gait and evaluation

Ivan Birch

COMPARISON OF GAIT

By this stage you should have a set of notes, a list of features of gait that characterise the gait of the figure in the questioned footage, based on a summary of the observations made from the various pieces of questioned footage examined, and a list of features of gait that characterise the gait of the subject in the reference footage, based on a summary of the observations made from the various pieces of reference footage examined. Your two lists of features of gait should have been derived completely independently of one another, and will provide all the information you are going to use for the next stage of the ACE-V process, comparison. Avoid going back to the footage while you are undertaking the comparison, as doing so could again introduce cognitive bias.

As with the preceding stages of the process, it is important that the approach taken to the comparison of features of gait is systematic, thorough, logical and unbiased. Your comparisons should be what is usually described as like for like. In gross terms, your comparison must be of features of gait seen to be exhibited by the figure in the questioned footage when walking, with those seen to be exhibited by the subject in the reference footage when walking, rather than, for example, comparing walking features of gait with running features of gait. Different locomotor activities involve different mechanical effects and can therefore result in different features of gait. However, in the forensic context you also need to consider the like for like nature of questioned and reference footage in terms of the factors that could affect gait and the factors that could affect your observation of the gait, including camera angle, frame rate, footwear, terrain etc. (see Chapter 7). It is important to remember that at this stage of the process you are not drawing conclusions, simply comparing the findings from two independent sets of analysis. Starting to draw conclusions as you work through the comparisons could bias your assessment of subsequent comparative observations.

For the purposes of comparison, the features of gait can now be categorised in three major groups of comparison outcomes:

i. **compatible features** These are features seen to be exhibited by both the figure in the questioned footage and the subject in the reference footage. These features, showing similarities between the gait of the figure in the questioned footage and that of the subject in the reference footage, will later in the process lend support to the proposition that the figure in the questioned footage is the subject in the reference footage.

ii. **incompatible features** These are features seen to be exhibited by either the figure in the questioned footage or the subject in the reference footage, which would preclude them from being the same person. These are rare. For a single feature of gait to be truly incompatible, it would have to be extremely unusual. An opposing combination of features seen in the questioned and reference footage is potentially more likely, but still rare. As an extreme example, if we observe that the figure in the questioned footage only has a left foot and that the subject in the reference footage only has a right foot, then we could consider the combination of the two observations to be incompatible. However, if we observe that the figure in the questioned footage only has a left foot, but the subject in the reference footage has both feet, we would need to be more cautious. Could the subject in the reference footage have a prosthetic foot? In other less extreme cases, you need to consider the possibility that the gait of the figure in the questioned footage has been significantly affected in some way between the collection of the questioned footage and the collection of the reference footage, perhaps by an injury, degenerative pathology or surgery. Many pieces of footage will contain, or have attached to them, the date on which they were captured, and it is often worth looking at the difference in the dates of capture of the questioned and reference footage. I have worked on cases where the

reference footage was captured up to two years after the questioned footage and anecdotally have heard of cases where the time intervals between the capture of compared materials have been even longer than that.
  iii. **features that differ but are not incompatible** Features that fall into this category can be subdivided into two sets:
     a. features seen to be exhibited by the figure in the questioned footage, but not by the subject in the reference footage
     b. features seen to be exhibited by the subject in the reference footage, but not by the figure in the questioned footage

These are the features that demonstrate differences between the gait of the figure in the questioned footage and that of the subject in the reference footage. These features will usually later lend support for the proposition that the figure in the questioned footage is not the subject in the reference footage, although in some cases the differences may be the result of factors you have already identified in the footage such as terrain or particular mechanical activities being undertaken by the figure/subject.

There is a fourth group of comparison outcomes, the features of gait for which a comparison could not be made, because it could not be determined whether or not the feature of gait was or was not exhibited by the figure in the questioned footage, or the subject in the reference footage, or both. Examples are the angulation of a joint at a particular point in the gait cycle being obscured from view by another part of the body, or the frame rate being too low to capture a particular movement on every step. In some casework you will have a number of such features. This is not the same as the absence of a feature of gait, which you have looked for and determined as not being present. That is a positive observation and will have been accommodated by the preceding three groups of comparison outcomes. The inability to make a comparison regarding features of gait adds uncertainty to the conclusions you make later in the process. Had you been able to make observations and then comparisons of these features, they may have supported either of the opposing propositions regarding the relationship between the figure in the questioned footage and the subject in the reference footage. Therefore, the more comparisons of features of gait that cannot be made, the greater the degree of uncertainty in terms of the conclusion that can be drawn from the comparison.

How you make the comparison, and subsequently how you present the comparison in your report, depends on the particular methodology you have employed to analyse the footage. There are currently two basic strategies commonly used: list and spreadsheet.

If you have produced lists of features of gait observed, you will now need to construct a series of lists under the three major group headings discussed above. If there are features that could not be compared, this should be acknowledged in your notes and in your report. When using this approach it is important to include all of the lists in the report, showing all of the similarities and differences. Omitting some of the lists will distort the representation of your findings. Remember that your responsibility as an expert witness is to the court, not under any circumstances to the commissioning agency. If you have been employed by the prosecution and you intentionally omit the list of features that differ, or if you have been employed by the defence and you intentionally omit the list of features that are compatible, you are wilfully failing to meet your responsibility as an expert witness (*The Criminal Procedure Rules The Criminal Practice Directions* October 2015).

If you have employed a spreadsheet approach to the observation of features of gait, such as that used by the Sheffield Features of Gait Tool (Birch et al. 2019) (see Appendix 1), all you need to do is construct a copy of the spreadsheet showing your summary findings from the questioned footage and your summary findings from the reference footage in adjacent columns. This can then be included in your report. When undertaking the comparison, you may find it helpful if in your notes you add a third column next to the questioned and reference footage summary columns in which you can note, perhaps by the colour of the cell, which features are compatible, which if any are incompatible, which differ, and which could not be compared. This will help you to gain an overall impression of the comparison.

One technique that some player software offers, that would seem to aid both comparison and the presentation of that comparison, is the ability to play two pieces of footage side by side, or in some cases to play one piece of footage superimposed over a second piece of footage (see Figures 8.1 and 8.2).

FIGURE 8.1 Images of the perpetrator, taken from the questioned footage, and of the suspect, taken from the reference footage, shown side by side.

FIGURE 8.2 Images of the perpetrator and the suspect, shown superimposed.

While this would seem to be a straightforward way of comparing the gait of two figures, I have yet to find any published research into the effects of cognitive bias in such a process. What has to be considered is what proportion of the population, when footage of them walking is superimposed upon the figure of interest, appears to a lay person to be the same person? I was once asked my opinion on the use of superimposed footage by a UK police force who were intending to use the technique as part of a television appeal for information regarding a murder. The footage showed the perpetrator walking across the field of view at some distance from the camera. Superimposed on this questioned footage was footage of the suspect taken from the same camera. The result was, from the police's point of view, a rather pleasing effect of the two images appearing to be the same person. The questioned footage had been submitted for use in height estimation, which had resulted in an estimation of the figure's height, with a possible error range of plus or minus a few centimetres. Allowing for the possible error in height, and the potential for an associated error in the width, what proportion of the local population would have yielded a similar effect? As soon as the superimposed images give the impression to the viewer of being of the same person, there is the potential for the introduction of cognitive bias. While an expert witness, with a thorough understanding of the possibility that the apparent similarity is the result of the technique, could take account of the potentially misleading nature of the use of the technique, a lay jury with no experience of gait analysis may not be able to exercise an appropriate level of caution in assigning evidential weight to the footage they have seen. The use of superimposed pieces of footage should therefore be used with extreme caution, particularly when presenting evidence in court where the potential limitations of the technique must be thoroughly explained.

EVALUATION

Having gained a thorough understanding of the similarities and differences between the features of gait that can be observed in the questioned and reference footage, the next step is to draw a conclusion from the comparison as to the evidential outcome, and how much probative value, in your opinion as an expert, can be assigned to the gait analysis evidence. As far as gait analysis is concerned, evidential reporting is, in the majority of cases, based on two opposing propositions. One is that the figure in the questioned footage is the subject in the reference footage. The other opposing proposition is that the figure in the questioned footage is not the subject in the reference footage. There are therefore three fundamental outcomes to evidential forensic gait analysis:

- the forensic gait analysis supports the proposition that the figure in the questioned footage is the subject in the reference footage
- the forensic gait analysis supports the proposition that the figure in the questioned footage is not the subject in the reference footage
- the forensic gait analysis evidence provides no assistance in addressing the issue

Arriving at an outcome involves the consideration of two elements of your work so far: the features of gait you have observed and noted, and the limitations of the footage you have identified. Barristers have in my experience simply added up the number of features of gait that are compatible and compared the total to the number that differ. This is, from a lay perspective, a logical but rather simplistic approach, the major flaw of which is that features of gait are not equally discriminatory. A missing lower limb is more discriminatory in most circumstances than a foot pointing outwards when weight bearing. On the basis of the published evidence, your education and your experience in gait analysis, you may consider some features, or combinations of features, of gait to occur less frequently in the population than others. On the basis of this collective knowledge, differences or similarities in some features, or combinations of features, of gait will carry more probative value than others. You also need to take into account the uncertainty added by the number of features of gait for which a comparison could not be made. As a result, drawing conclusions from your comparison of features of gait is considerably more complex than simply adding up the number of similarities and differences.

Once you have given consideration to the comparison of the features of gait, you then need to give consideration to the limitations of the footage you have identified. These will need to be clearly listed in your report. The limitations will attenuate the probative value you can place on your conclusion. For example, if the questioned footage showed a limited number of steps being taken by the figure, even if the features of gait observed from the questioned footage were a good match to those observed being exhibited by the subject in the reference footage, the fact that the comparison was based on the limited number of steps in the questioned footage introduces an element of uncertainty as to the degree to which the features of gait observed during those limited number of steps accurately represent the usual gait of the figure. The probative value of the comparison is attenuated by the limitation. The assignment of a probative value to the gait analysis evidence is therefore based on careful consideration of the strength of the comparison of the features of gait, and an equally careful consideration of the uncertainty introduced by the limitations.

The method by which the probative value of gait analysis evidence is typically expressed is based on the European Network of Forensic Science Institutes' scale of verbal expressions (2015). The origin of this scale is linked to the calculation of likelihood ratios, and in some areas of forensic practice the use of this scale is based on such calculations. This is not the case in forensic gait analysis, where the use of the scale is based on logic, but is an entirely opinion-based approach, informed by the gait analysis. This is an aspect of your conclusions that should be made very clear in your report and when giving evidence in court. The scale can be viewed as having, in effect, 11 levels, as shown in Figure 8.3.

The centre of the scale represents an outcome that the forensic gait analysis supports neither of the opposing propositions based on the combination of the features of gait that were seen to be compatible, seen to differ or could not be compared, and the limitations of the footage. This position is sometimes referred to as 'no evidence', a rather misleading term as there clearly is evidence, although the gait analysis evidence is, in this case, inconclusive. A more accurate descriptor is 'provides no assistance in addressing the issue'. Moving away from the centre of the scale are five levels of increasing probative value in each direction (limited or weak, moderate, moderately strong, strong, very strong) indicating increasing levels

very strong evidence to support	
strong evidence to support	the proposition that the figure in the
moderately strong evidence to support	questioned footage is the subject in the
moderate evidence to support	reference footage
limited evidence to support	
provides no assistance in addressing the issue	
limited evidence to support	
moderate evidence to support	the proposition that the figure in the
moderately strong evidence to support	questioned footage is not the subject in the
strong evidence to support	reference footage
very strong evidence to support	

FIGURE 8.3 Scale of verbal expressions of support.

of support for one or other of the opposing propositions. As has already been said, the scale has its origins in an association with the calculation of likelihood ratios, and the scale is designed to represent a logarithmic increase in probative value. It should therefore be remembered when using the scale that the selection of the next level away from the centre represents a tenfold increase in the suggested probative value of the gait analysis evidence. Gait analysis evidence is usually confined to the range between 'provides no assistance in addressing the issue' and 'moderately strong', with the occasional case yielding an outcome of 'strong'. Having said that, the *Forensic Gait Analysis: A Primer for Courts*, published by the Royal Society and the Royal Society of Edinburgh in the UK in 2017, suggested that "It is difficult to see how the use of any verbal equivalent other than 'weak support for the proposition' could be used to describe evidence provided by current approaches to forensic gait analysis" (The Royal Society and the Royal Society of Edinburgh 2017). As discussed by Nirenberg et al. (2018), although the intention of the document in providing consistent information regarding the use of gait analysis as evidence is wholly welcomed, there are aspects of the document and its conclusions that have elicited criticism within the forensic community. Whatever the outcome of the ensuing debate, the conclusions you arrive at from your analysis and comparison should be predicated on the gait analysis evidence alone, taking into account all aspects of the comparison of features of gait and the limitations of the footage, and the process you used to reach those conclusions should be transparent and demonstrable to the court. You have both a legal and moral responsibility to represent the evidence and your conclusions accurately and honestly, no matter who commissioned the work.

Figure 8.4 gives a summary of the work flow during the comparison and evaluation stages of the forensic gait analysis process.

The writing of your report will be considered in Chapter 9, but once your report is complete and you are happy with both the content and the presentation, the next step is to get your work verified by another appropriately qualified and experienced forensic gait analysis practitioner.

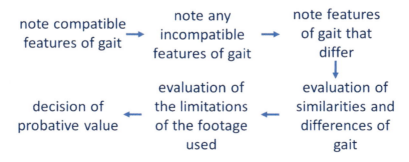

FIGURE 8.4 Summary of the comparison and evaluation stages of the forensic gait analysis process.

VERIFICATION

The first problem we have with verification is the term itself, which in the forensic arena is used to mean a number of things relating to both the process used and the outcome. Here we are using verification to mean a peer review of the casework by another competent and experienced forensic gait analyst. In the UK this is a requirement of the *Code of practice for forensic gait analysis* (2019) and therefore of the UK Forensic Science Regulator's *Codes of Practice and Conduct* (2017). All forensic gait analysis casework should be verified by someone who understands the process, has appropriate qualifications and has experience of forensic gait analysis, not just experience of gait analysis. The purpose of verification is not only to ensure that the gait analysis has been undertaken appropriately, but to ensure that the requirements of the forensic process have been met. Because forensic gait analysis is a subjective process, verification is crucial to establishing the validity of both the observations and the conclusions drawn.

You may of course be involved in the verification of forensic gait analysis in two ways: as the gait analyst whose work is being verified or as the verifier of someone else's casework. If you are the analyst whose work needs verifying, you will need to contact your prospective verifier during the early stages of the casework to check that they are happy to undertake the role and to allow them to plan time for the task. You will also need to ensure that the process to be used by the verifier follows the same principles of best practice as your own work on the case, and agree the cost of the verification, the completion date and the means by which the materials will be transferred and returned. Once you have completed the analysis and comparison, reached your conclusions, and written your report, you will need to send the following to the verifier:

a) a copy of the instruction together with the task being requested
b) detailed information regarding which pieces of footage show the figure/subject, and where in the footage in terms of timings the figure/subject is seen. If all or any of the footage shows multiple people, you will need to make it clear in your notes for the verifier which is the figure in the questioned footage and which is the subject in the reference footage.
c) the questioned footage
d) the reference footage
e) a copy of your report, including the features of gait observed

You will also need to make available a copy of your notes, particularly if the case involves multiple pieces of footage and therefore multiple lists of features of gait observed. Once the verifier has undertaken what is in essence a truncated version of the process already described in this chapter, and reviewed your findings and conclusions relative to their own, they should contact you to discuss any irregularities or differences of opinion. These could be with regard to the process, observations, terminology or the conclusions or probative value of the forensic gait analysis. If there are any such irregularities or differences of opinion, you as the analyst and the verifier must discuss them with the intention of reaching an agreement. Once agreement is reached, you should produce a final verified report which is then signed and dated by both you and the verifier. If there are outstanding areas of disagreement, these should be noted, in a way that is agreed by both the analyst and the verifier, and the notes kept with the other casework materials. If the disagreement is significant, it may be necessary for you and the verifier to agree to include an explanation of the areas of disagreement in the final verified report. This is a legitimate outcome, particularly in an area that is based on the subjective interpretation of data. If there are insurmountable disagreements, there is also the potential to employ a third examiner if absolutely necessary. It is important to remember that the verifier of your work is undertaking a peer review of the process used, the observations of gait, the comparisons and the conclusions. Their role is not to correct your spelling, typos and pagination. It is your responsibility as the report writer to ensure that the report meets the requirements of the legal system where you are practicing, and is professional in both content and appearance.

If you are acting as verifier for another analyst, your role is as the last port of call before the submission of evidence that could significantly affect the lives of other people. Your role is therefore a crucial part of the forensic quality assurance process. You should act as a critical colleague testing and challenging:

a) the process that has been used
b) the features of gait that have been observed
c) the comparison of the features of gait
d) the conclusions drawn
e) the probative values assigned to the conclusions

It is not your responsibility to be a proof reader.

In order to fulfil the role of verifier, you will need to undertake a truncated version of the process already described in this chapter. Do not read the report sent to you by the analyst until after you have completed your own analysis, comparison and evaluation of the footage, in order to reduce the potential for bias. Using the notes provided by the analyst regarding the timings and the figure/subject, you will need to review the questioned footage first and take notes of both the content and quality of the footage, as well as the features of gait observed. You will then need to do the same with the reference footage, before undertaking a comparison of the features of gait noted as being observed, as informed by the task requested by the commissioner. Finally, you will need to arrive at your own unbiased conclusions and assign a probative value to those conclusions based on both the comparison of the features of gait and the limitations of the footage.

Having completed all of these stages, you can then read the analyst's report, and assess how their work compares to yours, bearing in mind that they will have spent considerably longer undertaking the process than you have. If you agree with all aspects of their work and report, let them know and they can produce the final verified report for signing. If not, you will need to put aside enough time to discuss your findings and theirs with them with the objective of arriving at an agreement before the verified report can be produced. These discussions can be difficult, but they are part of the forensic process so you should not be reticent to say what you think, and if necessary disagree with the analyst's conclusions. As described above, if there are areas of disagreement that cannot be resolved, you will need to ensure that these are recorded and kept by both parties with the other casework materials, and if the disagreement is significant, it may be necessary to insist that an explanation of the areas of disagreement are included in the final verified report. The aim of forensic gait analysis is the production for the court of an objective and unbiased report based on an accurate and thorough assessment of the available gait evidence. As verifier, you are an essential part of the process that ensures that this is the outcome. Although not a common occurrence, as your name is on the report as verifier, it is possible that you could be called to give evidence in court and explain your agreement, or disagreement, with the analyst. It is therefore important that by the time you sign the verified report, you are entirely happy with the content, and would be able to explain and defend the report in court. It is also possible that you could be called to give evidence in court if the author of the report is, for a legitimate reason, unavailable.

Once the final verified report has been produced, it is your responsibility, as the gait analyst who undertook the casework, to recheck the report for typographical and content errors before signing and dating the report, getting the verifier to sign and date the report, and sending it to the commissioner together with a copy of your curriculum vitae. It may also be necessary to include other pieces of documentation requested by the commissioner, such as a witness statement form or self-declaration. Always ensure that you send reports by some form of secure and recorded delivery method that requires a signature upon receipt, and keep the proof of purchase with the casework documentation. Always use some form of rigid envelope for hard copies; you have just spent hours producing an evidence report, the last thing you want is for it to arrive folded or otherwise damaged in transit.

PAYMENT

The final stage of the process is to generate an invoice. As you will have already generated a quotation for the work, this should be relatively straightforward. Itemise each stage of the process, listing the time spent and the cost of each stage as well as the total. Remember to include the cost of postage of the report to the commissioner, and any other additional expenses, for which you will have to send the receipts with the invoice.

REFERENCES

Birch, Ivan, Maria Birch, Lucy Rutler, Sarah Brown, Libertad Rodriguez Burgos, Bert Otten, and Mickey Wiedemeijer. 2019. "The repeatability and reproducibility of the Sheffield Features of Gait Tool." *Science and Justice* 59(5):544–551.

Chartered Society of Forensic Sciences and College of Podiatry in association with the Forensic Science Regulator. 2019. *Code of practice for forensic gait analysis.* Issue 1. Birmingham: The Forensic Science Regulator.

European Network of Forensic Science Institutes. 2015. "ENFSI guideline for evaluative reporting in forensic science, strengthening the evaluation of forensic results across Europe (STEOFRAE)." Dublin. Accessed July 2019. http://enfsi.eu/wp-content/uploads/2016/09/m1_guideline.pdf

Forensic Science Regulator. 2017. *Codes of Practice and Conduct: For Forensic Science Providers and Practitioners in the Criminal Justice System.* Issue 4. Birmingham: The Forensic Science Regulator.

Nirenberg, Michael, Wesley Vernon, and Ivan Birch. 2018. "A review of the historical use and criticisms of gait analysis evidence." *Science and Justice* 58(4):292–298.

The Criminal Procedure Rules. The Criminal Practice Directions. October 2015. As amended April, October and November 2016, February, April, August, October and November 2017, April and October 2018 and April 2019. Accessed May 2019. www.justice.gov.uk/courts/procedure-rules/criminal/docs/2015/criminal-procedure-rules-practice-directions-april-2019.pdf.

The Royal Society and the Royal Society of Edinburgh. 2017. *Forensic Gait Analysis: A Primer for Courts.* London: The Royal Society.

9

Writing expert witness reports

PART 1: GENERAL PRINCIPLES, REQUIREMENTS AND PITFALLS

Roger Robson and Claire Gwinnett

An expert who is active in casework will be required at some point in their careers to present their evidence of opinion to the Criminal Justice System. This can be a daunting task, not only in providing testimony in court but also in the creation of an expert report which presents the results and interpretation in a clear and appropriate manner. Creating a well-presented, unbiased, helpful report that explains complex ideas and processes, in a language that is understood by the lay person, is not easy, but there is a plethora of advice, guidelines and published research that can be of help.

This section aims to introduce key principles of report writing, the general contents of a report, the legal requirements of reports, processes for the verification and submission of the report and potential pitfalls to avoid in report production. This guidance is not specific to the UK and as such we have refrained from specifically stating the obligations for an expert giving testimony in a UK court of law, although we have provided examples based on the UK where appropriate.

If you would like to delve more deeply into the writing, structure and impact of forensic reports, an excellent place to start is the 2000 paper by Evett et al., which was referenced in the National Research Council of the United States National Academy of Sciences (NAS) report, *Strengthening Forensic Science in the United States: A Path Forward* (National Research Council 2009).

KEY PRINCIPLES OF EXPERT REPORT WRITING

Forensic experts may be required to produce technical reports, provide written advice which assists an investigation, and usually later, a factual or evidential (opinion-based) report or statement. In fact, in legal terms, they serve the same purpose in that they all assist the triers of fact in reaching a decision. Initial reports may assist with the strategy of the investigation and may be considered non-*evidential*. In England and Wales for example, Level 1 Streamlined Forensic Reports (SFR1) summarise the expert's scientific facts but are not to be used as evidence in court. For the purposes of this chapter, we will focus on the opinion-based report which is served as evidence to the court and will refer to these as expert (witness) reports for consistency. It equally applies whether you, the expert, have received instruction from the prosecutor or the legal representative of the defendant.

An expert's report is required when the expert will be either, via the report or in person, presenting evidence of opinion in court. The report requires the expert's opinion to be voiced within some tight constraints: that the report reflects your area of expertise and within the limitations of the science, and that this is your scientific opinion and not your personal one.

Ultimately, the content of any expert report is dependent on the discipline and the case, but there are some fundamental principles to consider when creating a report. The first is to consider whether the report would be a satisfactory account of your work and findings in lieu of oral testimony. If the report cannot provide a clear stand-alone explanation of the evidence analysis, interpretation and conclusions, then it has already failed its key duty to the court. A benefit of a clearly written, well justified and robust report is that it is less likely to be challenged and the expert subsequently called to court, reducing costs to the criminal justice system.

To aid in the production of a clear and robust expert witness report, below is a list of some key principles to consider during their production.

1. Language and style of writing

Style of writing is one of the most important elements to get right when constructing an expert report. Scientific language can be a potential obstacle to the non-scientists reading the report, thus the language used must enable better understanding of the science in the report, not inhibit it. Although you are permitted to write in a relatively free style, it is always worth bearing in mind that your report will be read and scrutinised by individuals with varying levels of knowledge and education, which may range from scientists acting for the opposing legal team, defendants or appellants, members of the jury, the judge and members of the legal team. The term 'readability' is used when considering how easy a report is to understand by a non-scientist. How 'readable' your report is depends on your ability to be clear to all such parties and at the same time demonstrate to the court that you are an expert in this subject matter and are producing an opinion to assist the court in reaching the right decision.

What is acknowledged as good scientific writing by scientists is likely to have issues when being read by non-scientists. Bearing this in mind, there are four key features that have been identified as creating barriers for the non-scientist to their understanding of scientific reports. These include (Howes et al. 2014a):

- lexical density (using a high proportion of information-carrying words vs words used for a grammatical function)
- abstraction (the use of a large number of nouns to describe actions rather than using verbs as we would do in normal conversation)
- technicality (the use of specialist terms)
- authoritativeness (seen in the use of technical jargon and writing in the 3rd person and in the passive voice which is not a standard approach in ordinary English)

To understand the extent to which these features occur in your own report, the 'readability' of your report can be assessed in several ways. Howes et al. (2014a and 2014b) conducted studies on the readability of both glass and DNA expert reports and suggested a variety of evaluation methods, such as using a Flesch–Kincaid (FK) grade level, which provides an approximate number of years of US education required to read the text. A simple way to understand how readable your report is, is to check readability statistics in Microsoft Word®. The latest versions of Word allow you to select 'show readability statistics', which upon checking the document provides information that aims to improve clarity and conciseness. Readability statistics commonly include the number of sentences per paragraph and number of words per sentence. By viewing these statistics, you can identify if you are using overly long sentences and dense paragraphs of information. Although this chapter does not explicitly outline the different approaches used to assess readability, the suggestions below are based on ways to improve readability.

    i. Writing in first or third person:

        It is common practice to write in the first or third person. Writing in the first person can feel more natural when outlining your own opinion and expertise.

ii. Technical language:

Defining key terms and explaining complex scientific approaches and methods in a simple way is required if you wish for a non-expert to understand your report; this can be as part of a glossary, a footnote or clarified within the text. If you decide to use a glossary, refer to this early in the report and put terms in bold. All acronyms need defining the first time they are used, and it might be that these definitions need expanding upon to ensure the reader has a full understanding of their meaning and significance.

iii. Sentence length:

The use of very long sentences makes it more likely for the reader to miss the point. Break complex sentences down into two or three shorter sentences with overlapping points, i.e. restating information from previous sentences to retain flow (Howes et al. 2014a).

iv. Grammar:

These days we appear to be less hung up about grammar. However, your report needs to be readable and there should not be any glaring errors or misspellings, and no grammatical errors that change the way the reader interprets what you have written.

v. Use of generic report macros:

Whilst generic report macros are commonplace, be careful not to cut and paste irrelevant paragraphs, or worse, incorporate inaccurate text or reference to another defendant as this only looks clumsy at best.

vi. Be careful with your nouns and adjectives:

The legal team who cross-examine you are likely to focus on words they may not be happy with, or ones they hope will cause you to stumble. If you are a chemist for example, make sure you can instantly explain relatively simple familiar chemical terms such as compound or polymer. They may be words that you use daily, but better not to flaunt them around unnecessarily if you can't explain them in very simple terms.

vii. Be cautious with the use of descriptive words:

For example, "I had a very productive meeting with the Senior Investigator". That may appear helpful and an innocent comment until the defendant's Barrister enquires: "Why was it productive – what were you trying to do – find my client guilty?" You can soon find yourself in hot water in the witness box; try to stick to the facts: "I had a meeting with the Senior Investigator" covers the point admirably.

2. Impartiality and professional objectivity

Your conclusion needs to demonstrate that you have considered all likely scenarios that may explain your findings. Just because you have been instructed by the police or prosecutor, this does not equate to you writing a report that is favourable to them.

Similarly, if instructed by the defence team, you should avoid producing a report that reads as though you are attacking the expert instructed by the prosecutor. On this occasion, your role to the court is to review their work and ensure their approach is logical, the methods are robust, and opinions expressed are balanced. Be professionally critical and question decisions as necessary, but ultimately your report for the defence should be equally balanced as opposed to having been written in a style to appease the defence lawyer.

Furthermore, remember that the report is yours and no one else's. YOU will be the expert called to court; no-one else will be stood in the witness box with you, or be able to guide you. Therefore, do not allow yourself to be coached by a superior, nor rewrite your report in their style or include anything that you feel uncomfortable with via their instruction. Do not allow yourself to be coerced by a lawyer into changing the context of what you have said to the advantage of the prosecutor or defender.

3. Layout

The layout of the report should clearly define each section, be easy to navigate and have a logical order. The sections should flow and link together, avoiding 'surprise' information being included in a later section that was not referred to when needed earlier in the report.

Like any good report, expert or otherwise, headers, good diagrams and photographs are often useful. Headings and sub-headings should be consistent and concise but informative to the lay person. Consider using presenting headings in bold text but do not use all capital letters as the reader can no longer see the 'shapes' of the words, which decreases readability (Howes et al. 2014a). If you are using indentations to improve layout, ensure they are also consistently used.

Consider the use of white space in your report; text that is clustered together is harder on readers' eyes and more difficult to navigate. Maybe consider 1.5 or 2.0 line spacing to improve readability. Using spacings between paragraphs makes reports more appealing and reader friendly, so think about space management to improve the overall appearance of the report. If your eyes become tired just looking at the page, you need more white space.

The pages of the report should be numbered to ensure nothing is missing when served and it is worthwhile using the footer page counter to assist with this, e.g. page 1 of 50. Your report must always be signed and dated on the date it was submitted. A poorly ordered report can be confusing to the courts, so before starting, check the requirements and mandatory sections expected for a given country's reports, which will help in creating the initial structure.

Overall, remember that your report is an extension of your knowledge, skills, experience and work on a particular case and thus the responsibility of final proof-reading and checking is down to you. Upon submission, you should have confidence that it reflects your true scientific opinion in a manner that can be understood by a non-expert. To help provide this confidence, Box 1 gives some final thoughts to reflect upon before submission of your report.

BOX 9.1 SOME THOUGHTS FOR REPORT WRITING

1. **Do not disturb:** Put sufficient quiet time to one side to focus on writing.
2. **Readability:** Keep it simple and clear – a lay person needs to understand what you have written. Consider using a method of assessing readability.
3. **Seamless:** Use a style that flows – with a beginning (about you and what you have been asked to do), a middle (methods you've applied and the results) and a logical conclusion (expressing your carefully considered opinion based on the interpretation of the results).
4. **Attention to detail:** Include enough detail so that the courts fully understand what analysis was completed, the results generated and how you have come to the conclusions you have.
5. **Biased?** Ensure your style is independent. Stick to the facts and scientific analysis.
6. **Jargon free:** Extensive technical details are better kept separate in an appendix.
7. **Logical:** Explain how you have reached your opinion.
8. **Corrections:** Check your report and ensure what you have written is within your area of expertise.
9. **Page numbers:** Ensure all pages are numbered and you have included a statement noting total number of pages in the report.
10. **Signatures:** Check if there is a requirement in the country in which the evidence is being delivered to sign every page of the report (in many it is fine to just sign and date the statement noting number of pages in the report).
11. **Formatting:** Check that formatting, such as font, font size and margins, is consistent throughout the report.
12. **Double check:** A second independent check from a peer is always recommended.

GENERAL CONTENTS OF AN EXPERT WITNESS REPORT

There are many online resources that provide templates, guidelines and content lists of what should be present in an expert witness report. These can be helpful, but care should be taken that they are appropriate for the evidence you are presenting and also meet the requirements of the courts in your country. Your own organisation may have a template that they advise their employees to use, so check locally if this is the case. Even with guidelines, there may still be optional sections that can be included, for example, in England and Wales, the expert's occupation and age, the need to sign every page, and specific headings used for the report are not mandatory. The exact content will be dependent upon your instructions and the scope of the work but regardless of this, certain sections are generally always present. These are outlined below.

1. Declarations

It is likely that in most countries there is a mandatory set of declarations that will need to be included. It is important that you check with relevant governing or advisory bodies that you are including the correct information and wording, if this is stipulated. In England and Wales, declarations include the number of pages in the report, a Duty to the Court (stating that the expert has complied with, and will continue to comply with, their duties to the court, including unbiased, independent assistance within their expertise) and a Statement of Truth (which essentially states that the expert believes, to the best of their knowledge, that their statement is true). In England and Wales, this has very particular wording, which should not be paraphrased or re-worded. This is as follows:

> This statement (consisting of # pages) is true to the best of my knowledge and belief and I make it knowing that, if it is tendered in evidence, I shall be liable to prosecution if I have wilfully stated in it anything which I know to be false, or do not believe to be true.
>
> **Forensic Science Regulator 2017a**

It is worth checking if it is a requirement in your country for a declaration certificate to be included as an additional document.

2. Introduction

This includes your name, your role, qualifications and a short biography of your experience. For the latter, there is no longer a requirement to include a curriculum vitae unless specifically requested; the legal teams are usually happy with a maximum half page overview of your experience. Standard inclusions in this biography are the current post being held, types of expert examination you perform and the length of time you have performed these for. It might be that for highly specialised areas of expertise, relevant training should also be outlined, as well as more standard qualifications. You should also outline any limitations to your experience and whether anything discussed in the report is outside your expertise. The main aim of exploring limitations is for you to assure the court that you have the correct expertise and equally that the court understands any limitations of any methods undertaken. Keep the biography relevant to the task to which you have been instructed; there is little point making lengthy reference to, say, a PhD related to plant toxins, if your role as an expert for this case is specific to gait analysis. It is not wise to embellish your experience by using phrases such as "I was involved in the successful conviction of Mr Smith for his role in the brutal murder of his wife". The court will recognise you as an expert and will not generally be asking you to prove you are. Should there be any issues from either side as to your credibility to give evidence then such matters are mopped up before the trial. As such, these matters should be kept out of your report.

3. Instruction

This section includes the instructions put to you as the expert, by whom, when and what they were. Rather than use the instructor's words verbatim, you may need to rewrite them: "Are the hairs on the sofa from the assailant?" At this stage, one is not aware if this is the assailant or not; indeed, nor is one aware if a crime has been committed as it is an allegation. As such, you will be better sticking to the facts of the instruction, but rewriting it: "I was asked to examine the hairs recovered from the sofa (Exhibit A) and compare them with the reference head hair sample taken from Mr X (Exhibit B)".

Do you believe you have all the facts to be able to undertake your work in a thorough, conscientious, timely and professional manner? If not, you should state that and explain any shortfalls. If you consider that being supplied with additional facts (missing or new) that may come to light could change your opinion, then you should say so.

4. Background information

It is common practice for the instructing body to provide you with the necessary background information related to the alleged crime in order for you to make an informed opinion as to how your findings relate to the offence in question and the questions raised relate to the defendant. However, there is a cautionary note here in that you need to assure yourself you are not being drawn into thinking that a person is clearly the perpetrator and you need to ensure your evidence assists in a conviction. That is not your role. Nowadays it may be better to supply the expert with minimal background information, that way mitigating the potential risk of introducing bias – knowingly or otherwise. Ultimately, and however instructed, you must ensure you record all the information provided to you that you have relied on to produce the report, and that which was not used.

5. Samples supplied

In this section you will need to list the exhibits you have examined with clear descriptions and exhibit numbers. It is also good practice to refer to those you decided not to examine, perhaps with an explanation as to why you consider they would not affect your opinion either way. You should also include the date you took control of any exhibits and their origin, to assist the court in demonstrating a chain of continuity of said items.

6. Methods/analytical techniques used (aka forensic examination record)

The previous sections can be fairly self-explanatory to a lay person, but this section can be difficult to understand to a non-expert if not clearly written. Technical approaches, scientific terms and complex methodologies will need to be fully explained in a manner that makes it clear as to what was completed, in what order, and by whom. You should always stipulate how you proceeded with your examination, the methods on which you have relied, in what order they were applied and any precautionary measures you had to take. Any limitations or hindrances encountered at each stage should be described, including whether the test was destructive, should a further instruction to re-test emerge in due course. Be clear if you haven't done all the work yourself. If you had assistance, you should say who from, what their contribution was, and indeed, as you have relied on their contribution, you need to say who they are and their relevant qualifications and experience. This information may be better in its own section so that it is clearly highlighted to the court.

If the methods are recognised nationally or internationally and/or your laboratory has this method accredited, it can be mentioned here. Equally, clearly state any other methods referred to which are not covered by accreditation and any mitigation you have taken to ensure the method has been robustly applied. Rather than going into the technical detail of each method here, if it is required, the appendix may be a more suitable place to include this. To avoid interrupting the flow of the report, any additional details about the method that the expert wishes to be declared may be also be included in the appendices.

7. Results

This section should contain the data and observations that your opinion has been based upon. For the courts to understand how you have come to these conclusions, well-presented data is crucial. Raw data is not acceptable in this section; the data needs to be in a format that allows even the most complex to be clearly understood with appropriate explanation. Transforming raw data into appropriate formats to show the comparison of samples, the outcomes of statistical tests and any outliers, should be well thought out and justified. Transparency is key here; the tests applied, and any data transformations, should be explicitly stated, including the use of any existing or new datasets/databases. If the latter is used to aid in the interpretation of your results, the size and scope of the datasets along with details of how they have been generated, compiled and queried should be included too. This allows for the courts to understand whether the datasets are valid and applicable for use in the case. If the instruction has led you or your team to create swathes of data, rather than break up the report, it may read better if you include analytical data sheets as an appendix and summarise the results in an easily digested table/summary within the results section. If tables, photographs, graphs or diagrams are used, ensure they are labelled with appropriate descriptions to guide the court in understanding what they comprise. Simple things like checking any units of measurement are provided consistently in the correct SI units, and making sure table column headings provide suitable information, can be forgotten during proof-reading. A good question to consider when reviewing your results section is "Would I understand this table/graph/result if I were not an expert in this field?" If the answer is probably not, then consider what can be changed or included to aid understanding.

Finally, when checking your results section, ensure that the results included represent the entirety of your work, such that there is sufficient quantity to justify your opinion and that it also includes results that might undermine your opinion. It is also good practice to state that all of the results are available upon request.

8. Discussion

A good discussion section sheds light on the results, making them comprehensible to the lay person. The discussion should firstly describe the results in a way that translates the data that was presented in the results section into useable conclusions. Key points should be highlighted, and any trends and comparisons made between samples noted. The discussion should then explain these results; what do these results actually mean? How do these results help answer the questions being asked in your instructions? The interpretation approach utilised should be fully explained and justified.

9. Conclusion/evaluation

Arguably this is the most important section of all, and may be the section that gets read first by the courts and defines the rest of the report. The correct wording and structure of this section are imperative for your opinion to be understood in relation to the question being addressed. You should strive for an unbiased, balanced and succinctly written conclusion. Try to demonstrate a balanced approach by discussing realistic alternatives, ending with your overall conclusion and your overall opinion, based on a level of support if you use one.

Many experts these days, when asked to express an opinion, follow a simple alternative proposition model, e.g. whether or not Mr Jones was driving the vehicle at the time of the crash. This approach is easily understood by the courts and helps to avoid contradictory opinions.

Remember, your role is not to prosecute, so stick to your role to look at supportive scientific evidence which may refute an allegation in just the same way. It is not your role to link your conclusions with the innocence or guilt of those at trial, nor indeed whether a crime has been committed. You should therefore avoid referring to the deceased as the 'murder victim' or someone being tried for breaking and entry as the 'intruder', for example.

If you are aware that there is likely to be a range of opinions from your peers, it is wise to explain what that range is and justify why you have chosen a particular point on a scale. These scales of evidential value are often subjective, and it is always best to be open in such matters.

Likewise, if you are aware of anything that would detract from the opinion put forward, you should state this, such as the uncertainty of a measurement or a method applied that has limitations.

It is simpler to avoid the inclusion of complex statistics in this section, such as Bayesian inference, which if used, should be available separately through disclosure.

Remember that the conclusion is the section that the courts are likely to be most interested in and they will refer to it to quickly understand what your ultimate opinion and findings are, so double check that your opinion is clear, cannot be misunderstood and is not masked with complex language or complicated jargon. Your conclusion should be consistent with the rest of your report. This seems obvious, but it can sometimes be tempting to introduce something additional here that is not relevant and confuses matters.

10. References

You are likely to want to support your conclusions and the approaches you have used by referring to accepted published methods and information. The reference list allows the courts to view the range and types of sources of information that you have used. The reference section provides the courts with a list of primary sources which have provided background context to the sample type, the analysis techniques used, data which has been used to help interpret your findings and any underpinning ideas and theories which you have relied upon. Naturally, any literature you have relied upon to reach your conclusion must be listed. Robust, peer-reviewed and up-to-date primary sources, such as internationally recognised journal articles, are excellent references to use. You can even reference unpublished papers here but only if they are of particular relevance. Remember, the courts do not need to see pages and pages of references listing every paper on your expert topic, but do want to see what you have relied upon to reach your opinion.

11. Appendices

As stated in earlier sections, the appendix can be a very useful location to include supplementary information that although not mandatory, may help the court by providing additional information. Ensure that the appendices are referred to in the report, labelled clearly, are ordered appropriately and are formatted consistently with the layout of the main report. Like other sections in the report, they should be easy to navigate.

Legal requirements of an expert report

It is imperative that an expert witness report abides by any particular legal requirements for your country and courts. Failing to do so may lead to a report not being accepted and the evidence contained not being useable in the case. All experts would agree that this is not a favourable situation to be in. Although some general considerations are outlined below, if in doubt, always seek advice from your instructing body or national guidance. For example, for England and Wales, the Forensic Science Regulator's *Expert Report Guidance* (2017a), which supports the *Codes of Practice and Conduct for Forensic Science Providers and Practitioners to the Criminal Justice System* (Forensic Science Regulator 2017b) and outlines the requirements imposed by certain prosecuting authorities.

As noted earlier, most countries expect an expert witness to fall in line by stating some form of declaration which includes your duty to the court. Your primary role as an expert witness is to express your independent expert opinion which you have reached based on the information that has been provided to you. You may have instigated tests or re-tested samples or undertaken a review of another expert's work as part of your instruction.

As the expert, you may give sworn evidence to a court of law. There is a very clear expectation from the court that you will act independently and impartially to others. Even if employed by the police, the judge will soon discount your evidence if you show undue favouritism and they have the power to discredit you.

The court will also expect to see you are appropriately competent to pass an opinion. There may be legal debate at times as to whether some experts, despite many years' experience, are deemed competent on a specific subject matter. It is the judge's final decision whether the court wishes to hear your evidence. Bearing this in mind, it is worthwhile clearly stating in your report which of the facts detailed are within your own knowledge and equally making clear those that are not.

Some countries, including the UK, now insist on including a Statement of Truth within your declaration where you declare that you have written the report in the belief that its contents are truthful to the best of your knowledge. In addition, you have an understanding that, if your report is put forward as evidence, you are liable to be prosecuted yourself if you have wilfully stated anything you knew to be false.

As such, there is a legal requirement to include your name, qualifications and experience. It is not a legal requirement to state your occupation, although to include such is often helpful. Oddly, most courts ask for your age (but only if under 18), but the authors are not aware of any experts under that age that this would apply to! You also have an obligation to sign and date your report.

Relatively recently, with the introduction of data protection rules in some countries, it is also advisable to make some reference to disclosure within your declaration, ensuring you can demonstrate you have control of the exhibits and have dealt with any data generated from them which relates to identity; what is stored and what has/will be destroyed. It may be simpler just to say you have read and followed your national guidance on such matters.

Verification and Report Submission

Most legal teams now accept an electronic pdf report with your e-signature. If not then it is your responsibility to ensure a paper copy reaches the instructing party swiftly and safely, such as posting the report in an A4 secure envelope using tracked and signed for delivery.

Prior to submission, it is wise to go through a final checklist to ensure nothing is missing and all information is correct and presented appropriately. Box 2 provides a checklist that can be used prior to final submission, but you may want to create your own, including additional points which you want to be reminded of.

If after submission of your report, you notice an error or further information comes to light that affects your interpretation, an additional statement may be submitted to the instructing body. You should make clear any changes to your interpretation and conclusions and link these directly to your original report. You should not adjust your original report but be very clear as to what errors are present.

Potential Pitfalls in an Expert Witness Report

There are many ways that an expert can unintentionally create a report that is misleading or unhelpful to the court. Table 9.1 outlines some of these ways (this is certainly not an exhaustive list) and possible ways to avoid them.

> ### BOX 9.2   FINAL CHECKLIST FOR REPORT WRITING
>
> 1. Check it yourself for clarity. Have you clearly answered the question(s) posed?
> 2. Final proof-read including dates, times, names, addresses, exhibit identifiers all correct.

3. Checked by designated verifier(s) – both the analyses and opinions expressed.
4. Check the report complies with national guidelines, legal rules and obligations placed on you as an expert witness.
5. Don't forget to sign and date your final report.
6. Keep a copy with your case notes, together with any drafts and evidence of the peer review and checks on your report – whether paper and/or electronic copies.

TABLE 9.1
Common Pitfalls and Possible Ways to Avoid them

Common Pitfall	How to Avoid
Unintentional bias writing, intentional bias – being unduly led by the legal team	Step back from the severity of the crime, the suspect, the pushy investigator and your personal morals. Be aware of factors that may cause bias, such as irrelevant contextual information (Stammers and Bunn 2015). Your role is to assist the court with the scientific facts and be independent to both the prosecutor and defender.
Misleading the court that you have done all the work yourself	Ensure you refer to any assistance you have had in performing the work and reaching your conclusion.
The use of jargon – even relatively simple scientific nouns and adjectives should be considered carefully as you may well be asked to clarify them during cross-examination	Remember your audience and write in lay person's terminology. If necessary, footnotes can be used and/or a glossary of terms included in the appendices.
Drifting away from your own area of expertise	Do not stray from your area of competence, an area that you can evidence if need be.
Not being provided with all the facts	Include a statement to say you have relied only on the information supplied and you may wish to revise your opinion should new facts come to light.
Over-complicating your conclusion	Keep it simple and punchy. All your deliberations explaining how you reached your opinion should have been explained elsewhere prior to the conclusion. If your conclusion is long-winded you may wish to clarify a single point or two by creating a summary, which should be placed towards the front of the report.

REFERENCES

Evett, I. W., Graham Jackson, J. A. Lambert, and S. McCrossan. 2000. "The impact of the principles of evidence interpretation on the structure and content of statements." *Science and Justice* 40(4): 233–239.

Forensic Science Regulator. 2017a. *Expert Report Guidance FSR-G-200.* Issue 1. Birmingham: Forensic Science Regulator.

Forensic Science Regulator. 2017b. *Codes of Practice and Conduct for Forensic Science Providers and Practitioners in the Criminal Justice System.* Issue 4. Birmingham: Forensic Science Regulator.

Howes, Loene M., Paul K. Kirkbride, Sally F. Kelty, Roberta Julian, and Nenagh Kemp. 2014a. "The readability of expert reports for non-scientist report-users: Reports of forensic comparison of glass." *Forensic Science International* 236: 54–66.

Howes, Loene M., Roberta Julian, Sally F. Kelty, Nenagh Kemp, and Paul K. Kirkbride. 2014b. "The readability of expert reports for non-scientist report-users: Reports of DNA analysis." *Forensic Science International* 237: 7–18.

National Research Council. 2009. *Strengthening Forensic Science in the United States: A Path Forward.* Washington, DC: National Academies Press.

Stammers, Sophie, and Sarah Bunn. 2015. "Unintentional bias in forensic investigation." POSTbrief, in Houses of Parliament Parliamentary Office of Science and Technology. Accessed July 2019. https://researchbriefings.parliament.uk/ResearchBriefing/Summary/POST-PB-0015

PART 2: WRITING FORENSIC GAIT ANALYSIS REPORTS

Ivan Birch

Having gained an understanding of the general principles of writing a forensic report, you can now draw on what has been said in the previous chapters and consider the particular areas that need to be covered in a forensic gait analysis report. As has already been pointed out in Part 1 of this chapter, the format of your report will of course largely depend on the jurisdiction in which you are working, but below are a few pointers as to likely content.

Name and relevant qualifications and experience

Provide enough information for the court to gain an insight into the grounds on which you are claiming to be a gait analysis expert witness. However, this is not a curriculum vitae (which can be appended to your report when it is submitted), so do not clutter the section up with unnecessary detail in an attempt to appear more qualified than you actually are.

The task requested and undertaken

State the task that was agreed and who requested that you undertook the work. Ensure that the question that was posed to you in the task is clearly stated. When you give your conclusion later in the report it can then be couched as the answer to this question. If appropriate in the jurisdiction in which you are working, it is useful to include at this early stage in the report the current prerequisite statement to the effect that 'in producing the report, you understand your duty to the Court and have complied with that duty and will continue to comply with that duty', as is used in the UK.

A list of exhibits

Provide a clear and unambiguous list of all the DVDs, data sticks, hard drives and any other media and materials that you were provided with for use in forensic gait analysis. List exactly the labelling on the media, including any spelling errors, and the exhibit number if it has been provided. If there are only a limited number of files on the media, you may wish to list them here as well. If there are a lot of files, it is often better to list them in an appendix, which you can refer the reader to. The advantage of using an appendix is that you can then list detailed information about the various pieces of footage such as the player used to analyse them, and which pieces were used and which were not, without cluttering the main body of the report.

An introduction to forensic gait analysis

For many of the readers, your report will be their first engagement with forensic gait analysis and indeed gait analysis in general. You therefore need to ensure that the reader has some background knowledge and understanding of gait analysis, its history, its uses, the research evidence base underpinning its use, and its development and limitations of its use in the forensic context. A good introduction to this area of professional practice at this stage of the report will help everyone involved to understand what can and cannot be done with gait analysis, and provide context and justification for your analysis and conclusions.

How you examined the exhibits

Much of the work of the forensic gait analyst is done on a computer, which is therefore a fundamental part of the process. State the type and specification of the computer, processor and monitors you used to do the work. If you have not included this information in an appendix with the list of footage, you need to say what software was used to play the footage. Give a brief description of how you went about viewing the footage, such as: 'all footage utilised for the purposes of forensic gait analysis was viewed at normal and slow speeds, with areas of particular interest also being examined frame by frame'.

In this section you can also provide information regarding the quality of the footage, such as resolution, lighting and frame rate, where it was captured, and the number and position of the cameras relative to the figure/subject.

An important point to note is that all the questioned footage was viewed, analysed and notes taken before you repeated the process for the reference footage.

How the analysis and comparisons were undertaken

Describe how you recorded the features of gait from the footage, and how you compared the features seen in the questioned footage with those seen in the reference footage. As described in Chapter 8, you need to explain that you considered compatible features, incompatible features and features that differed but were not incompatible. You need to provide enough information for the reader to follow your methodology precisely so that they can understand how and why you came to your conclusions. If you have used a tool to record the features of gait, then you can include a modified version in this section to show the comparative findings. If you have not used a tool, then you will need to list the features of gait observed that fall into each of the categories. Do not omit any feature of gait observed, as to do so could mislead the reader.

Your evaluation and conclusions

To ensure that the reader understands the context of your conclusions, list the limitations of the footage clearly and fully early on in this section, together with any other limitations under which the work has been conducted. The nature of your conclusion will of course depend on the task you were set and possibly on the jurisdiction in which you are working, but in most cases your conclusion will be a level of support or rejection for one of two opposing propositions as described in Chapter 8. State the two propositions and then your conclusion, if appropriate assigning a probative value to your findings using an accepted scale. If there are required statements relating to the findings and conclusions of forensic practice in the jurisdiction in which you are working, be sure that they are included.

It is also important that you make clear the nature of your findings. In most cases forensic gait analysis is an opinion-based practice, so you need to say so in the report. A statement along the lines of 'this is an opinion-based conclusion, and is not predicated on numerical data or statistical calculation' will make this clear.

Declarations

As described in Part 1 of this chapter, there will almost certainly be a long list of required statements that must be included in all expert witness reports in the jurisdiction in which you are working. Make sure they are all included, and check regularly for amendments to the list. After the declarations you will need to sign and date the report, as will the verifier.

Appendices

These could include additional background information on gait analysis and its use in the forensic context, a full list of the footage you were given, which footage was used and which was not, which software was used to play the footage, an explanation of the scale of probative value you have used and the research that underpins its use, and a reference list. While the appendices are a legitimate and important part of your report, there is a word of warning associated with their use. Based on personal experience, in the UK there is a tendency for the criminal justice system to save money on photocopying by not including your appendices in copies given to other members of the Court. It is therefore important to remember that your appendices should include supporting information, but not information that is fundamental to the process you have used and your conclusions.

10

Presenting gait evidence in court

Ivan Birch

Giving evidence at trial may be the most common aspect of being an expert witness that stops more people from engaging in the work than any other. The role of the expert witness is to assist the court in answering the ultimate question by providing an objective opinion based on knowledge and experience. This would seem to be a relatively straightforward task. However, the nature of a jury-based adversarial legal system makes this far from straightforward, and indeed sometimes deeply unpleasant. The purpose of the adversarial system is to test the resilience and credibility of the evidence presented, but this inevitably includes testing the resilience and credibility of the witness as well. There are of course other types of legal systems, where the evidence is considered by a panel of judges who are able to gain background information about the area of work in question and pose informed questions to the witnesses. For an expert witness this is probably a better option, removing at least to some extent the necessity to provide a simplified explanation of the evidence to a jury of lay people. Whatever the type of legal system that you end up giving evidence in, there are a number of basic things that should be considered. This chapter will provide information that may help those less experienced in presenting evidence in court. It does not seek to provide all the information you will need, or reiterate information readily available from other sources (Daeid 2010, Smith and Bond 2014, Bowen 2017), but to provide guidance on areas that you may not have thought of, based on the personal experience of the author.

CONSIDER GIVING EVIDENCE AS YOU WRITE YOUR NOTES AND REPORT

The possibility of a trial can seem a long way off when you are writing your notes and report, but a little forethought here can smooth your experience in the witness box. You have presumably already read the section on report writing (Chapter 9), so you understand the importance of accuracy and attention to detail. Read your report through carefully, then after removing any names or identifying information to ensure confidentiality, get a trusted critic to proof read it, then read it again yourself before you send it for verification. The verifier may find things that warrant change, so when your report comes back from verification, do the changes and then read it through yet again before submission to the commissioning agency. A trusted colleague of mine once said that if you don't want to have to answer a difficult question in the witness box, don't let it be asked. What he meant was that getting everything exactly right in your report, clearly, concisely and correctly punctuated, will reduce the opportunities for criticism under cross-examination. When you are being cross-examined, the barrister[1] may attempt to undermine the credibility of your evidence, and the credibility of you as a witness. People's lives will be affected by the trial, and testing the evidence and the expert witness is appropriate. Spelling mistakes and typos are a perfect starting point. The same is true for your notes. You are required to keep detailed contemporaneous notes of your work, and it is likely that if you are called to give evidence at a trial that you will be asked to submit a copy of your notes for the court. So again, make sure that your notes are correct, clear and concise, and that you can identify every piece of footage from your notes. If you are using a features of gait tool as well as long hand notes, be very sure before submitting your report that the notes, the entries on the tool and your report all match in every detail. Investing another hour or two at this stage will be well worth the effort when you are in the witness box. If you find mistakes in your notes or in the tool you used, note and acknowledge them, but do not change your notes or the entries in the tool.

Well organised commissioning teams will ask you for dates on which you cannot attend court to give evidence either at the start of the process or on report submission. This will allow you to plan for the trial well in advance. But this is not always the case, and the shortest notice of giving evidence I experienced was 24 hours. The dates of trials also get changed, sometimes repeatedly, and even if the trial dates remain unchanged, the days of the trial on which you are required to attend may change, often at short notice. Your notes, your features of gait tool and your report should of course all be finished by the time you submit the report, so be sure that at report submission they are all in a state in which you would be happy to submit them to the court.

NOTIFICATION OF REQUIREMENT TO GIVE EVIDENCE

When you first get the letter informing you that you are required at court, there are a number of options regarding the content. The most likely is that the date and time at which you have been instructed to arrive is the start of the trial, rather than the time you are needed. The letter is usually produced by the administration team of the court, and a phone call to your contact in the commissioning team is therefore nearly always necessary to ensure that you arrive when you are needed rather than at the start of the trial. For a short trial, arriving at the beginning may be fine, but for a long trial of several weeks you could have been instructed to attend several days or even weeks before you are actually needed. It is worth mentioning here that criminal trials are very fluid events with unexpected twists and turns that can speed things up (rarely) and slow things down (frequently). So you need to keep your travel and accommodation arrangements as flexible as possible. Booking your hotel at a discounted non-refundable rate may seem like a good idea, but when the day they want you to give evidence has changed two or three times in the preceding week, it can ultimately be expensive.

PRE-TRIAL MEETING

A pre-trial meeting with the commissioning team is an excellent strategy, but one that rarely gets employed. One of the best trial experiences I had happened because we had a pre-trial meeting several months before the trial. It allowed me to familiarise myself with the barrister and the police officers involved, and it allowed them to familiarise themselves with the gait evidence. Most importantly, Most importantly, it facilitated the organization of the logistics of presenting the gait evidence in court, what equipment was required and who was going to operate the equipment. The equipment available in the court to play and watch the footage will vary. I have encountered DVD players that will not play computer files and computers that will not play DVDs, and on a number of occasions have only been able to give evidence because I took my laptop with me with the evidence already loaded. On one occasion I had to leave the laptop with the court after giving evidence so that the jury could use it to view the evidence during their deliberations. The equipment available to view the footage varied similarly from one screen carefully positioned on the opposite side of the court room from the jury and just below a window, to one high definition monitor for each member of the jury, the judge, the legal teams and me. A short discussion with the commissioning team regarding the presentation of your evidence can avoid some nasty shocks when you arrive at court.

PREPARATION FOR THE TRIAL

The week before the trial at the latest, you should be preparing for giving evidence. View the footage, read your notes, and read your report carefully and repeatedly. At trial there will be some obvious early questions that you can be prepared to the answer. For example, what is gait analysis? Having well prepared answers to these relatively straightforward questions will allow you to settle down, not that you will ever be relaxed in the witness box, and show the court that you are well informed and articulate. So prepare answers to obvious questions. This will include re-reading any references that you used during the evaluation of your findings. Put yourself in the position of the barrister on the other side: where would you start if you wanted to challenge the gait analysis report? Identify likely questions, and prepare and practice the answers out loud. This is particularly important if you do not regularly speak in public. There is a big difference between having an answer in your head, and getting that answer coherently out of your mouth. The barristers may well have scrutinised things you have written, or things you have said in court at previous trials. You must be consistent, and if there are apparent inconsistencies you need to be able to explain the reasons for these, such as more recent research or increased knowledge. In court you will only be questioned on what you have written in your report, but what you have written in your report may provide opportunities for wider questions. Make both electronic and paper copies of your report, your notes, any features of gait tool you have used, correspondence regarding the case and any other relevant records ready to take with you to court. If you can, get copies of the footage onto your secured laptop as well, just in case. A phone call to your contact on the commissioning team during the week before the trial is always worthwhile, to confirm timings and equipment availability and operation.

If you are ill on the day you are required to attend the trail, you will need to contact the court and/or the commissioning team. You can be excused if there is a genuine reason, and trying to give evidence when you cannot perform at your best could be seen a failure to meet your responsibilities to the court as an expert witness. If you are genuinely not able to attend, the verifier of your work may be asked to attend, or the presentation of your evidence may be moved to another day.

Finally, of course, you have got to arrange travel and accommodation.

GETTING TO COURT

Arriving at court with time to spare is good practice, but requires some careful planning. In the UK, morning sessions usually start about 10:00 and afternoon sessions about 14:00. Having established when you are actually needed, you then need to factor in time for a number of things before you can determine when to arrive. You need to allow plenty of time for travel problems, delayed trains, traffic jams etc., as a matter of course. You are probably going to feel stressed enough without the background worry of arriving late to give evidence. You will have been instructed to arrive at court for the start of either the morning session or the afternoon session. If you have been instructed to arrive for a morning session, have you got enough time to get to the court from your home on the morning safely, or are you going to have to travel the day before and find accommodation? If you need to find accommodation, my advice is find a hotel close enough to the court for a short walk in the morning and remove all doubt about getting there on time. Bear in mind that there may well be a limit on how much reimbursement you will receive from the court and/or commissioner, so five-star hotels and first-class travel may well be largely at your own expense.

SITE VISITS

When undertaking your analysis of the footage, you will probably have investigated the location of the footage capture using software such as Google Maps, to make sure that there are no facets of the terrain that are not obvious from the footage, such as slopes. An advantage of arriving in the vicinity of the court early is that it may give you a final opportunity to visit the place where the footage was captured. You may of course already have done this as part of the analysis, but if not, if you have any doubts about aspects of the terrain, a site visit can be very useful. However, site visits can also be problematic. The footage may have been captured by privately owned cameras, and the owners may not appreciate you looking at their property or being reminded of the crime, particularly if the footage was not willingly provided. Always let the commissioning agency know that you are intending to undertake a site visit, explaining why you think it will be useful or essential, and they will be able to forewarn you of any issues. If the site visit becomes part of your testimony, you will of course need to inform the opposing side in the trial that it has taken place and provide them with your notes. If you undertake a site visit, always ensure that you have appropriate identification and letters of permission if appropriate. If you are in any doubt regarding your personal safety at the site, arrange to go with a police officer or members of the commissioning team.

AT THE COURT

As we've already discussed, a pre-trial meeting is a good idea, but in many cases the best you are going to get is 10 minutes with the barrister before you give evidence. This is not ideal for a number of reasons. It is likely that the barrister has only had the information regarding the case for a relatively short time and therefore they are unlikely to have a full understanding of the work you do, what your evidence looks like and therefore what is going to be required in court when you give evidence. Until this meeting, they may also not have fully appreciated the role of your evidence in the trial, and occasionally this meeting will result in the presentation of your evidence being delayed or rescheduled. Although they will probably have anticipated that you are going to need to play the footage in court, it is possible that the technical challenges of doing so may not have become fully apparent. These not only include the available equipment but also who is going to operate the equipment and how are you as the expert going to communicate with the operator during the presentation of your evidence. This is sometimes the point at which when you have been commissioned by the prosecution that they spring on you that the defence have also employed a gait analyst who has also produced a report that you now have no chance of reading before giving evidence. Whether you have had a pre-trial meeting or not, you need to arrive at court with enough time to deal with the inevitable challenges.

Court buildings rarely open before 09:00, giving you about an hour to get any preparation done before the day's business begins, but the court rooms themselves often remain locked until after 09:30 unless someone asks for them to be opened. You therefore need to have arranged to meet someone from the commissioning team at a designated time specifically for the purpose of preparation. The hour between 09:00 and 10:00 will at least give you some time to prepare, but if you are unsure about the technical issues that may present themselves, or the evidence is particularly complex, then you may need to arrive at court and meet the commissioning team the afternoon before the day you are due to give evidence. The team will of course already be involved in the trial and will have limited availability, but arriving the afternoon before will give time for you to meet the team when they finish for the day or during recesses in the court's business. It will also give you the chance to get into the court where you will be giving evidence and stand in the witness box to see what it feels like. Is there enough surface space in the witness box for you to lay your notes out? Will you need two pairs of spectacles in order to read your notes and see the jury? Where are you standing in relation to the person operating the equipment? The layout of most court rooms is in general similar, but not the same, particularly in the case of some of the older British court rooms (Figure 10.1). The best case scenario is that you get an opportunity to rehearse playing the footage with the person who will be operating the equipment when you give evidence.

If you have been instructed to arrive for the afternoon session, my advice would be to get there as early on the day as possible so that the preparations discussed above can take place before the trial starts for the day, during recesses and during the lunch break. The rule of thumb is: expect there to be problems with playing the footage.

Even when you have arrived at court and met with the commissioning team, do not be surprised if you do not get to give your evidence on the day as planned. Criminal trials are complex and prone to organisational adaptation. My advice is if you have travelled some distance to get to court, take what you need for more than one day and an overnight stay.

FIGURE 10.1 A traditional English court room. This is not now typical of your working environment at a trial, but the author has given evidence in court rooms very similar to this one.

GIVING EVIDENCE

There are many texts and courses available for you to read up on how to give evidence once you get into the witness box, so I have no intention of repeating what has been written. What I will do is report a few things that I and my colleagues have learnt. The jury and the judge will have very clear expectations about what an expert witness will look like. You may not like their expectations, but you should make every effort to meet them. The following passage is taken from "The Criminal Courts, Criminal Law and Evidence. An Introduction for Forensic Scientists", a document written by Dr Ann Priston OBE:

> In a survey in the USA the 'ideal' forensic expert in court was envisaged as preferably male, in his mid 40's, wearing a dark three-piece suit and carrying a briefcase. He may wear glasses and has short hair. He is neat, intelligent and dignified looking; calm, professional and serious, confident and controlled. If the expert were a woman she would be in her late thirties, wear a dark skirted suit, her hair either short or put up and she would carry a briefcase and pocket book. She must look neat and intelligent and respond to questions calmly and confidently. Unlike her male counterpart, she is allowed to be a little nervous. She must be pleasant but at the same time, serious, honest and professional (Tanton 1979).

154 Chapter 10. Presenting gait evidence in court

One must not take this too seriously, but must remember that appearance and demeanour are important aspects of the expert's presentation.

The survey is dated and nowadays shockingly sexist and ageist, but the message is as important today as it was in 1979 (Figure 10.2).

Look business-like and smart in dark clothing and always wear an appropriate tie if you are male. What is happening in that court room will change people's lives dramatically. You have a role to play in that process, but you are not central to it and there are no advantages to be gained from dressing in an inappropriate or rebellious fashion, either for the process as a whole or for the credibility of your evidence. On a practical note, make sure you know where you can leave your belongings securely stored. Walking to the witness box carrying your coat, briefcase and wheelie suitcase will do little for your credibility. Take only what you will need to the witness box, and leave the rest somewhere safe. This raises the question of what to take to the witness box with you. You will be provided with a copy of your report by the court. Unfortunately, my experience has been that this copy is usually of very poor quality and therefore difficult to read. Taking your own copy with you will add an element of familiarity. Always take a copy of your notes, including any features of gait tool you have used, to the witness box with you. You should

FIGURE 10.2 We may have moved on from this demography in the court room, but some expectations regarding the appearance and behaviour of an expert witness remain.

ask permission from the judge if you have to refer to them, but have them there ready. And remember to take your reading spectacles with you. I have already alluded to the fact that you will be working at different focal lengths when you are in the witness box, so be prepared. After a few trials I had a pair of bifocal spectacles made specifically for giving evidence, so I could read my report and see the jury without changing spectacles.

Various courses are available to help you develop your court room skills, but be sure that any course you are planning on signing up for will support the development of skills for criminal trials. Civil trials can be very different. As before, there are a few key points to share, gained from first-hand experience. When giving evidence, speak clearly and do not rush. You need to be completely in control of what is coming out of your mouth. Plant your feet pointing towards the jury who you will be addressing at all times. This will help you to keep eye contact with them, and will reduce the impulse to turn and face the barristers. Usual rules of politeness would dictate that you face the person asking the question and address the answers to them. This is not how it works in court. Every question is asked for the benefit of the jury, so you do not have to make eye contact with the questioner, just the jury. The only possible exception is if the judge asks a question which requires you to answer them directly, in which case make eye contact with them as you answer. There is another aspect to not looking directly at the barristers, particularly during cross-examination. In an adversarial system, one of the purposes of cross-examination can be to shake the jury's faith in you and your expert evidence, and pantomime-esque body language, facial expressions and the throwing up of hands in mock exasperation are not unknown. Keeping your attention firmly on the jury will help to reduce the effect of any histrionics on the part of the legal teams on your ability to fulfil your role as an expert witness effectively.

Your time in the witness box will be divided into a number of sections. After you have taken the oath or affirmation, you will be asked to give some basic details about yourself, your qualifications and your expertise. This is your first opportunity to establish your credibility as an expert and why the court should listen to you. Answer factually without being egotistical, but this is not a time for undue modesty. This will be followed by what is referred to as examination or evidence-in-chief, during which the barrister for the commissioning agency will question you and lead you through your report. This is followed by cross-examination, during which the barrister for the opposing side will question you. In general terms, the difference between these two parts of giving evidence is that during evidence-in-chief the questions will tend to be open questions that allow you to expand on the answer, whereas during cross-examination the questions will tend to be closed questions that allow no room for expansion. During evidence-in-chief, your questioner is trying to make the most of your evidence in support of their case. The jury will not have seen your report and this is your opportunity to explain what you did, why you did it and what you found. During cross-examination, your questioner is trying to undermine the value of your evidence. This latter process is, in the experience of the authors, often undertaken by using one of two strategies. The first is to expand on a particular aspect of your evidence, drawing comment and agreement from you as the advocate carefully works their way to the very edge of your area of expertise and then slightly beyond. As an expert witness you can only offer opinions on matters that fall within your area of expertise, and this strategy seeks to encourage you to comment on matters beyond your area of expertise, for which you will then be openly criticised. This can undermine the credibility of all your evidence. The second strategy is to do the opposite, drilling deeper and deeper into one particular detail of your evidence, the end point of which is a question to which you either do not know the answer, or to which there is no answer. In either case, the advocate will make a meal out of pointing out that this so-called expert cannot answer a simple basic question on their area of expertise. After cross-examination, the barrister for the commissioning agency will have the opportunity to ask any further questions, or clarify and points. The bottom line when giving evidence is of course always tell the truth, but always acknowledge the limits of your knowledge, never become confrontational or argumentative (barristers will always be better at arguing than you) and never offer an opinion on anything that is outside of your area of expertise.

USING STILL IMAGES AS EVIDENCE

Still images taken from video footage have an important role in casework. They can be used to help identify the figure/subject in question, or demonstrate the location of a piece of footage. They may be a necessary way of demonstrating an aspect of a feature of gait. But they are not in my opinion and in a forensic context a good way of demonstrating a feature of gait to the court. What a still image lacks is the context of the frames that were captured immediately before and after. This lack of context can be, and in my experience has been, used to mislead the court deliberately. For example, if the defence expert has reported that the gait of the figure in the questioned footage is characterised by the right foot pointing markedly outwards and the prosecution expert produces as evidence a still image of the figure's right foot pointing inwards, they have introduced an element of doubt in the defence evidence. However, that still image could have been taken from a piece of footage where the figure is not exhibiting their usual mid step gait, for example when turning or even standing. Another example might be if the prosecution expert has reported that the subject in the reference footage tends to raise their left forefoot markedly immediately prior to heel strike, and the defence expert submits a still image of the isolated left foot showing the forefoot close to the ground. The still image could have been taken from any part of the gait cycle, not necessarily immediately prior to heel strike. So still images can be helpful, but they can also be very misleading. If you are demonstrating a feature of gait to the court, the court should not only see the feature of gait, but also the context in which the feature of gait occurred. Playing the footage at full speed, at reduced speed and frame by frame is in my opinion the best way to achieve this and meet your responsibility as an expert witness. If the opposing side asks you to comment on a still image from a single frame of footage, ask to see the context in which the still image was captured. What can be seen in the frames before and after the still image?

AFTER GIVING EVIDENCE

When the court has finished with you, you will be officially dismissed by the judge, and you can leave the witness box, remembering to take all your belongings with you. On some occasions, the commissioning team will ask you to assist them in cross-examining the opposition's expert, so will then join them in court while the other expert gives their evidence. If this is not the case, you are free to join the public gallery to observe the remainder of the trial. Once you have finished your official role in proceedings, I have found it useful to do two things if possible. The first is to speak to the commissioning team, which will enable you to get a feel for how your evidence has contributed to the trial. The second is to speak to the opposing barrister who cross-examined you. This may seem a little unusual, particularly if they have just given you a particularly hard time during cross-examination, but you will usually find that out of the court room they are in fact human, and very happy to have a conversation. Both of these simple things are opportunities for informed feedback and comment on your performance and your evidence that shouldn't be missed. They are also opportunities to network with members of the legal professions who may become clients in the future.

After you have left the court, you will often get little in the way of contact from the commissioning agency. This is not unreasonable as the team will very likely have moved straight onto the next case and trial. If you don't hear from them, contact them after a reasonable time period, perhaps four weeks, and ask about the outcome of the trial and sentencing and keep a note of these outcomes. Potential commissioners of work often want to know what you have done before and the contact details of people they can ask about your work. This is also information that you might be required to provide when applying to be listed on various professional registers and databases. After a trial you should reflect on your performance as an expert witness and identify what went well and more importantly what did not go so well. In my experience this is best achieved in conjunction with a debrief with another forensic professional with whom you can be open and honest, and who will understand the context.

NOTE

1. The term used to describe the person undertaking this role varies depending on the legal system. For example, in the US it would be lawyer or attorney.

REFERENCES

Bowen, Robin T. 2017. *Ethics and the Practice of Forensic Science*. Boca Raton, FL: CRC Press.

Daeid, Niamh Nic. 2010. *Fifty Years of Forensic Science: A Commentary*. Chichester: John Wiley & Sons.

Smith, Lisa, and John Bond. 2014. *Criminal Justice and Forensic Science: A Multidisciplinary Introduction*. London: Macmillan International Higher Education.

Tanton, R. L. 1979. "Jury preconceptions and their effect on expert scientific testimony." *Journal of Forensic Sciences* 24(3):681–691.

11

Psychology of perceptual error in forensic practice

Liam Satchell

As humans, we have an impressive array of senses at our disposal, but our success at navigating the world requires selective ignoring of information. The world is a noisy place and it would be impossible to perceive everything around us, from all senses, at all times. We can refocus our attention so that we are not, for example, consciously focussing on the sensation of our clothes pushing onto us. We can wear watches or have clocks that make a routine ticking sound, but we often do not 'hear' this constant noise. It is only when our senses are bored (when we're sat quietly or trying to get to sleep) that we become aware of the information we regularly tune out.

SENSE AND NON-SENSE: THE LIMITS TO HUMAN PERCEPTION

We 'selectively attend' to the sensory information that is most relevant to our current activity. Typically, this means paying attention to our sense of vision. Sight is one of the more dominant senses and we have built a world of signage and technology that requires visual attention. We often prefer vision over other senses, 'correcting' what we hear or feel to match what we see. An example is known as the 'McGurk effect'. In this classic psychological experiment, a silent video of someone saying "fa" can be paired with the audio of someone saying "ba" and the listener will 'hear' the person saying "fa" (McGurk & MacDonald 1976). The McGurk effect shows how easily our eyes can override our other senses. This effect is so instinctual and difficult to overcome that even those who know about the McGurk effect and prepare themselves for the phenomenon still wrongly hear "fa". By prioritising vision, we can regulate the information we receive to our overburdened senses. However, trusting vision above all other senses can blind us to the making of mistakes in what we see.

Vision itself is overburdened. The sense of sight is constantly exposed to a huge breadth of information. The human total visual field is 200 degrees (Hughes 1977). This means that, when looking directly ahead, our eyes detect visual information in an array all the way to just about each shoulder. This offers a tremendous range of sight and is advantageous for spotting anyone creeping up on us. However, this is an overwhelming amount of visual information for us to be consciously aware of at all times. We manage this by focussing our perceptual awareness on particular aspects of the environment. We have only limited attention for the whole of our full visual field unless something changes, for example if someone approaches us from the side. This focussed awareness makes us great at 'natural perception', such as planning a path into the distance or examining an object in detail. But fixation and passing awareness of our whole visual field are designed for active world exploration, where we can update and refine our perception through interaction. If something looks like a steep incline, we can change our viewing position to get an alternative view and update our perception. This ecological approach to perception is less suited to the requirements of modern life. We are forced to passively observe content inside videos, photographs or images. This means that we cannot interact with what we are watching, we cannot experiment with the presented material and refine our perception. In many regards, the way we have evolved to cope with the difficulties of perception makes accurately retaining information from video or photographs a real challenge. It is difficult to manage the quality of our perceptual attention when interacting with these media – to the extent that we can miss 'obvious' things in our environment.

Classically this was demonstrated in research by Simons & Chabris (1999), who studied 'inattentional blindness'. A sample of participants were shown video footage of students, dressed in a white or a black t-shirt, passing a basketball. Observers were asked to count the number of times the basketball was passed to a member of the white or black team. This task requires dedicated attention to the passing of the ball and is challenging, even with the relatively slow pace of the video. In fact, this task requires such focus that people will continue to count the passes of the ball and not notice a person dressed in a gorilla costume walk across to the centre of the video, turn to camera, beat its chest and then leave the screen. This highly unusual event goes unnoticed by a surprising number of observers. When people were told to watch the white team pass the ball to each other, only 42% of the sample reported noticing anything unusual in the video. When participants were told to pay attention to those in the black team (and so might deliberately pay attention to a black gorilla), one in five participants missed the gorilla in the basketball video. This finding of 'inattentional blindness' has been demonstrated many times with many different variations of video. The gorilla study itself was based on earlier work where someone would carry an umbrella, largely unnoticed, through a group passing a basketball (Neisser 1979). 'Blindness' of this type is to be expected, and even required, when our experience of the world is noisy and rich in information. With all that is calling for our attention, it is easy to miss aspects of our world that are not. The gorilla and umbrella examples are comical exaggerations of this effect, but there are important consequences to inattentional blindness in forensic-related contexts. When following a suspect across different CCTV cameras, inattentional blindness can lead to not noticing that the person on screen has changed (Davies & Hine 2007). The same process can also lead to mistaken eyewitness identification (Nelson et al. 2011).

It is easy to think that this effect would 'not happen to me' or that high-stakes observation would lower the risk of inattentional blindness. Unfortunately, a broad range of experts are vulnerable to these errors. Even expert basketball players, who are routinely well-practised at identifying important aspects of dynamic basketball scenes, often demonstrated basketball-related inattentional blindness (Furley, Memmert, & Heller 2010). In other contexts, expert radiologists, whose job it is to notice minor abnormalities in a scan, did not notice a gorilla 48 times the size of a normal lung-nodule inserted into an evaluation material (Drew, Võ, & Wolfe 2013). In some ways, experts are more vulnerable to such errors. The pressures and high stakes of professional work can require dedicated cognitive resources, enabling inattentional blindness.

There are further opportunities for mistakes in perception. Assuming a situation where we correctly observe and attend to something in the world, we then need to make sense of it. Sense-making involves piecing together the information in the world to make sense relative to ourselves. However, observing the world afresh each day would require an overwhelming amount of effort. Instead, we have a catalogue of 'shortcuts' in our thinking. Psychologists refer to these quick and easy ways of sense-making as 'heuristics' (Gigerenzer & Gaissmaier 2011). For example, we can have a 'door handle' heuristic. When we see a door with an additional bar around hand height, we can perceive an opportunity to pull the door open. This heuristic is highly useful and is successfully used every day. However, this heuristic is far from perfect. Like all shortcuts in thinking, it creates opportunities for mistakes, such as when we encounter push doors with a pull-typical handle or pull doors with an inviting push-panel. In this case, our sense-making shortcuts are misled by environmental information and the direction (and amount) of force we use with the door can be inappropriate. Many heuristics, like the door handle heuristic, are more often than not effective, and so receive little scrutiny when they do go wrong.

The McGurk effect is an example of an effective heuristic going wrong. However, it is more often than not useful. There are many benefits to using vision to help us if we mishear someone in conversation. Further, in daily life it is unlikely that we are presented with sound and visual information that is contradictory like the videos which we use in the artificial McGurk experiment. So, in most situations, the heuristic error shown in the McGurk effect is functional, supporting everyday perception. Similarly, inattentional blindness is also functional in most cases, as a way to survive the noisiness of the world. As heuristics are frequently successful, we usually have limited awareness of our own mental shortcuts. As such, we often do not re-evaluate those thinking shortcuts that may be leading us to make mistakes. The class of heuristics that lead us to misperceive the world is known as 'cognitive biases'.

FOCUS ON FORENSIC SCIENCE

All of the above limitations on our senses, our attention and our sense-making show that our perception of the world is by necessity subjective – unique to each of us, depending on what we filter out and see as important. However, the purpose of any type of forensic evaluation is to provide courts with objective evidence to help legal decision making. The forensic analyst is, therefore, in a challenging position. Their activity requires them to go above and beyond the normal parameters of perception. Human errors, in the form of cognitive biases, are ever present in science. Even the most 'scientific' forensic analysis requires human interpretation and reporting. It is due to this that there has been a move to study cognitive biases in forensic science more fully.

Cognitive bias in forensic science has been highlighted with some key case studies. Notably, concerns about bias in fingerprint matching were raised after a significant misidentification in identifying a suspect in the 2004 Madrid Bombings (Koen & Bowers 2017). The Spanish National Police (SNP) lifted a fingerprint from a bag containing detonation devices in a van near the bombing site. The fingerprints were sent to the US Federal Bureau of Investigation (FBI) for matching. The FBI suggested 20 possible matches, and on May 15th 2004 performed background checks into the matches. Part of this investigation revealed that one potential match, Brandon Mayfield, a former US army officer and practicing lawyer, had recently converted to Islam. This background information put Mayfield under heightened suspicion. On May 17th, three FBI agents concluded that Mayfield's fingerprints were a match to those on the bag. The SNP came to a different conclusion and later arrested and convicted another suspect, demonstrating Mayfield to be innocent and the fingerprints not to be a match. After legal action, the US government formally issued an apology to Mayfield and paid compensation to him and his family (Koen & Bowers 2017).

Psychological research on the topic of fingerprint matching grew in the aftermath of this case. The research showed, much like the Madrid case study, that analysts are vulnerable to relying on information beyond the target samples to draw judgments. One experimental study on fingerprint matching provided mock analysts with case information before being asked to decide if two non-matching fingerprints were a match. Four alternative case information packets were manipulated so that both the type of crime committed and the suspected perpetrator varied. Participants were more likely to find a 'match' for the two non-matching prints if the crime was higher stakes (child abduction as opposed to burglary) and if the suspect description fit the stereotypical offender (Smalarz et al. 2016). Fingerprint matching is vulnerable to bias as the analysts' task is difficult. Matching two abstract images is beyond the realm of normal perceptual experience. Even playing 'spot the difference' with two photographs with only subtle differences can be a real challenge, a type of 'change blindness'. So, it follows that analysts would unwittingly draw on heuristic knowledge information to attempt to improve their accuracy in the task. It is the case that many of these heuristics are wrong and prejudicial stereotypes can hinder performance.

This is not unique to fingerprint matching, with research showing that human factors can bias handwriting examination (Found & Ganas 2013) and forensic anthropology (Nakhaeizadeh, Dror, & Morgan 2014). Dror (2018) summarises the research on the psychology of forensic science at large and notes that contextual case information can lead to errors across multiple areas of forensic science practice. So widespread are the concerns about errors in forensic analysis that the United States National Academy of Sciences released a report stating that forensic sciences "rely on human interpretation of what could be tainted by error" (National Research Council 2009), and the UK's House of Lords Science and Technology Committee launched an inquiry into the status of forensic science in the UK (Forensic Science Inquiry 2018).

The research showing issues with bias in forensic analysis has also suggested strategies to mitigate and control bias. Principally, this has been to encourage better division between casework at large and forensic analysis. Those engaged with the interpretation of forensic science should be kept apart from as many details of the particular case as possible. They should not be exposed to wider case details that could lure them into using their heuristics to influence their interpretation. Dror & Cole (2010) suggested that analysts should not be informed about potential suspects or eyewitness accounts, and just evaluate the evidence presented to them. This has been productively deployed in the case of firearms examinations, where accuracy in identifying matches improved when domain-irrelevant case information was not presented to examiners (Mattijssen et al. 2016). This is not a perfect solution, however, as it can be the case that some forensic analysis necessitates wider casework information. An example of this is forensic blood pattern analysis, where interpreting a particular stain requires information about the wider events of the crime (Osborne et al. 2016). Forensic gait analysis has similar challenges. It is difficult to attempt to compare gaits without being unintentionally exposed to other case details.

FORENSIC GAIT ANALYSIS – POTENTIAL PITFALLS

To date, there has not been a detailed examination of cognitive bias in forensic gait analysis. However, from wider knowledge on bias in forensic science and the psychology associated with human gait, there are three potential heuristics that could influence an analyst's work.

First, the case information biases detailed above also apply to forensic gait analysis. Comparing gait requires the same forced-passive perception as other image-based comparison in forensic analysis. Comparison is often reliant on CCTV footage which is lower quality than everyday vision. Analysts, like anyone working in impoverished contexts, are vulnerable to using heuristics to 'fill in' the information they cannot perceive. Further, security cameras are often placed higher up than normal eye level and the way in which analysts have to observe the gait is from an atypical perspective. If gait analysts are working from limited visual information, it is even more likely that cognitive biases could affect judgment. Thus, vulnerability to unintentionally using case detail heuristics when coming to a judgment is heightened.

The most obvious case information that may bias an analyst is the knowledge that the individual in the footage is a likely suspect. Effectively, analysts are being asked a leading question by the interested parties. This information alone increases the likelihood of finding a match between questioned footage and reference footage. This particular subset of bias is known as 'confirmation bias', where we often look for evidence to confirm previously held beliefs. If the analyst's task is perceived as the verification of a suspected match, a match a reasonable person might assume has been suggested as a result of a thorough investigation, then the analyst might be more likely to confirm a match. Further, if the request comes from the criminal justice system, it may be the case that implicitly, the analyst understands that finding a match may serve the 'common good'. Alternatively, if an analyst's services are retained by a defence, they could believe that there is a reasonable argument to find no match. These processes are largely pre-conscious and not necessarily how an analyst would verbalise their decision making. These effects may also be small, affecting very few cases, but it is the sort of effect that trainees and novices might be vulnerable to in particular.

The second challenge for a gait analyst is that the comparing activity is dependent on observing the questioned and reference person's body. As mentioned above in the case of the Madrid Bombing and the research on target stereotyping on fingerprint matching, it can be challenging to separate personal implicit biases about people from comparing or matching activities. Everyone holds implicit or 'unconscious' biases about those who belong to our outgroup. In many subtle ways we disfavour those who are not part of our identity of 'us'. This could be based on an individual's perceived ethnicity, gender, sexuality, law obedience or other criteria where we may define people as like 'us' or not. For a fingerprint analyst, we can remove case details to help protect them from their own natural biases. For a gait analyst this is more challenging, as appearance and context are nested inside target footage. As such, it is important to be aware of mistakes we may make due to what we can see in a person's appearance.

Thirdly, gait itself provides opportunities for our heuristics to guide thinking. Gait, unlike fingerprints or blood patterns, has inherent social information. How someone moves acts as a social signal to those around them. Decades of research have shown that gait can inform us about the sex of another person (Runeson & Frykholm 1983), their sexuality (Johnson et al. 2007), their vulnerability (Gunns, Johnston, & Hudson 2002, Johnston et al. 2004) and their general aggression (Satchell et al. 2017, 2018). An analyst may be presented with footage where the target person just looks aggressive from the gait they are told to focus on. They look like the kind of person who would engage with crime. Further, gait is variable and can act as a signal for someone's intentions. Whilst individuals have distinctive gaits, it is the case that in times of heightened 'affective states' (i.e. anger, lethargy, happiness etc.) our gait changes (Birch, Birch, & Bray 2016). In fact, gait change can be seen before individuals move to engage with a violent event (Troscianko et al. 2004). Whilst it could be considered a 'good thing' for an analyst to note, for example violent intentionality in gait, this biases attempts at comparing gait. Given the social and intention information present in gait, the analyst has to be wary of using information beyond the movement parameters when reaching a judgment.

FORENSIC GAIT ANALYSIS – STRATEGIES FOR MINIMISING BIAS

The above hypothesised risks for forensic gait analysis could be mitigated with strategies that could minimise the emergence of heuristic-based perception. Firstly, in line with recommendations for forensic science at large, the work of forensic gait analysts should be as removed as possible from case information (Dror 2018). Analysts should be provided with target footage that is sufficient for analysis (i.e. involving multiple gait cycles) but should be removed as much as possible from any crime-relevant details. Details about the wider case, the nature of the crime, witness accounts and any physical evidence should not be presented to gait analysts.

Secondly, as a field-wide standard, gait analysts should engage in self-review (periodically reviewing one's previous cases), peer-review and adversarial peer-review for most, if not all, cases. An analyst's reasoning should stand up when another person with similar expertise, but different cognitive biases, reviews the same footage. If possible, this process should be blind to the case details, the original analyst's identity and any previous judgment reached (so as to avoid bias against particular peers). This anonymous review system also helps to protect against confirmation heuristics by considering both similarities and differences in gait.

Thirdly, structured processes and checklists should be implemented for gait analysis, to increase the objectivity of the analysis process (Birch et al. 2019). Such processes and checklists should be evidence-based and standardised. Any attempt at increasing the objectivity and non-natural observation of target footage should be encouraged. The less natural the approach to reaching a judgment is, the less likely the analyst is to be reliant on heuristics and bias. A systematic process would not only help the reliability of an individual analyst's findings, but would also help make the process of peer-review more robust.

Fourthly, there should be more empirical research on the work of forensic gait analysts. The above suggestions are reasonable strategies against bias drawn from wider forensic literature. However, gait analysis is a unique task in terms of forensic analysis. The fact that there is no way to remove the intentionality and social information present in gait, means that gait itself could be biasing analysts. It is not yet known to what extent. Further work between psychologists and analysts would help better understand the vulnerabilities for bias and counter-strategies that are unique to forensic gait analysis.

CONCLUSION

Perception is challenging. We live in a complicated world with an array of sensory experiences that could overwhelm us if not for our selective attention. Selective attention may help us explore the world, but it makes objective forensic analysis challenging. Research has shown that everyday shortcuts in perception, called 'heuristics', can mislead us when we are engaging in less-natural tasks. Specifically, when we are forced to experience the world passively, our 'cognitive biases' can lead us to erroneous perceptions that we cannot correct through interaction. Cognitive biases have been well studied in some areas of forensic science, and have led to UK and US forensic science regulators publicly declaring caution about subjectivity in forensic analysis. Forensic gait analysis is vulnerable to these biases and more. In particular, forensic gait analysis has the challenge of never being able to remove information about the target person from the footage being used for comparison. Gait itself has social value and it is challenging to separate the role of an abstract analyst from the everyday human perception of gait as valuable. Potential counter-strategies exist though, such as distance from case information, peer-review and systematic processes and checklists.

REFERENCES

Birch, I., T. Birch, and D. Bray. 2016. 'The Identification of Emotions from Gait'. *Science and Justice* 56(5): 351–6. doi: 10.1016/j.scijus.2016.05.006.

Birch, Ivan, Maria Birch, Lucy Rutler, Sarah Brown, Libertad Rodriguez Burgos, Bert Otten, and Mickey Wiedemeijer. 2019. 'The Repeatability and Reproducibility of the Sheffield Features of Gait Tool'. *Science and Justice* 59(5): 544–551.

Davies, Graham, and Sarah Hine. 2007. 'Change Blindness and Eyewitness Testimony'. *Journal of Psychology* 141(4): 423–34. doi: 10.3200/JRLP.141.4.423-434.

Drew, Trafton, Melissa L.-H. Võ, and Jeremy M. Wolfe. 2013. 'The Invisible Gorilla Strikes Again: Sustained Inattentional Blindness in Expert Observers'. *Psychological Science* 24(9): 1848–53. doi: 10.1177/0956797613479386.

Dror, Itiel E., and Simon A. Cole. 2010. 'The Vision in "Blind" Justice: Expert Perception, Judgment, and Visual Cognition in Forensic Pattern Recognition'. *Psychonomic Bulletin and Review* 17(2): 161–67. doi: 10.3758/PBR.17.2.161.

Dror, Itiel E. 2018. 'Biases in Forensic Experts'. *Science* 360(6386): 243. doi: 10.1126/science.aat8443.

Forensic Science Inquiry – Select Committee on Science and Technology: House of Lords. 2018. *Forensic Science: Oral Evidence*, Parlimentlive.tv, 9 October 2018. Accessed October 2018. www.parliament.uk/business/committees/committees-a-z/lords-select/science-and-technology-committee/inquiries/parliament-2017/forensic-science/

Found, Bryan, and John Ganas. 2013. 'The Management of Domain Irrelevant Context Information in Forensic Handwriting Examination Casework'. *Science and Justice* 53(2): 154–58. doi: 10.1016/j.scijus.2012.10.004.

Furley, Philip, Daniel Memmert, and Christian Heller. 2010. 'The Dark Side of Visual Awareness in Sport: Inattentional Blindness in a Real-World Basketball Task'. *Attention, Perception, and Psychophysics* 72(5): 1327–37. doi: 10.3758/APP.72.5.1327.

Gigerenzer, Gerd, and Wolfgang Gaissmaier. 2011. 'Heuristic Decision Making'. *Annual Review of Psychology* 62(1): 451–82. doi: 10.1146/annurev-psych-120709-145346.

Gunns, Rebekah E., Lucy Johnston, and Stephen M. Hudson. 2002. 'Victim Selection and Kinematics: A Point-Light Investigation of Vulnerability to Attack'. *Journal of Nonverbal Behavior* 26(3): 129–58. doi: 10.1023/A:1020744915533.

Hughes, A. 1977. 'The Topography of Vision in Mammals of Contrasting Life Style: Comparative Optics and Retinal Organization'. In: Hughes, A. (ed.) *The Visual System in Vertebrate*, 613–756. Berlin, Heidelberg: Springer.

Johnson, Kerri L., Simone Gill, Victoria Reichman, and Louis G. Tassinary. 2007. 'Swagger, Sway, and Sexuality: Judging Sexual Orientation from Body Motion and Morphology'. *Journal of Personality and Social Psychology* 93(3): 321–34. doi: 10.1037/0022-3514.93.3.321.

Johnston, Lucy, Stephen M. Hudson, Michael J. Richardson, Rebekah E. Gunns, and Megan Garner. 2004. 'Changing Kinematics as a Means of Reducing Vulnerability to Physical Attack'. *Journal of Applied Social Psychology* 34(3): 514–37. doi: 10.1111/j.1559-1816.2004.tb02559.x.

Koen, Wendy J., and C. Michael Bowers, eds. 2017. *Forensic Science Reform: Protecting the Innocent*. Amsterdam/Boston: Elsevier/AP, Academic Press is an imprint of Elsevier.

Mattijssen, E. J. A. T., W. Kerkhoff, C. E. H. Berger, I. E. Dror, and R. D. Stoel. 2016. 'Implementing Context Information Management in Forensic Casework: Minimizing Contextual Bias in Firearms Examination'. *Science and Justice* 56(2): 113–22. doi: 10.1016/j.scijus.2015.11.004.

McGurk, Harry, and John MacDonald. 1976. 'Hearing Lips and Seeing Voices'. *Nature* 264(5588): 746–48. doi: 10.1038/264746a0.

Nakhaeizadeh, Sherry, Itiel E. Dror, and Ruth M. Morgan. 2014. 'Cognitive Bias in Forensic Anthropology: Visual Assessment of Skeletal Remains Is Susceptible to Confirmation Bias'. *Science and Justice* 54(3): 208–14. doi: 10.1016/j.scijus.2013.11.003.

National Research Council (U.S.), ed. 2009. *Strengthening Forensic Science in the United States: A Path Forward*. Washington, DC: National Academies Press.

Neisser, U. 1979. 'The control of information pickup in selective looking'. In: A. D. Pick (ed.) *Perception and Its Development: A Tribute to Eleanor J. Gibson*. Hillsdale, NJ: Lawrence Erlbaum Associates, 201–219.

Nelson, Kally J., Cara Laney, Fowler Nicci Bowman, Eric D. Knowles, Deborah Davis, and Elizabeth F. Loftus. 2011. 'Change Blindness Can Cause Mistaken Eyewitness Identification'. *Legal and Criminological Psychology* 16(1): 62–74. doi: 10.1348/135532509X482625.

Osborne, Nikola K. P., Michael C. Taylor, Matthew Healey, and Rachel Zajac. 2016. 'Bloodstain Pattern Classification: Accuracy, Effect of Contextual Information and the Role of Analyst Characteristics'. *Science and Justice* 56(2): 123–28. doi: 10.1016/j.scijus.2015.12.005.

Runeson, Sverker, and Gunilla Frykholm. 1983. 'Kinematic Specification of Dynamics as an Informational Basis for Person-and-Action Perception: Expectation, Gender Recognition, and Deceptive Intention'. *Journal of Experimental Psychology: General* 112(4): 585–615. doi: 10.1037/0096-3445.112.4.585.

Satchell, Liam, Paul Morris, Chris Mills, Liam O'Reilly, Paul Marshman, and Lucy Akehurst. 2017. 'Evidence of Big Five and Aggressive Personalities in Gait Biomechanics'. *Journal of Nonverbal Behavior* 41(1): 35–44. doi: 10.1007/s10919-016-0240-1.

Satchell, Liam, Paul Morris, Lucy Akehurst, and Ed Morrison. 2018. 'Can Judgments of Threat Reflect an Approaching Person's Trait Aggression?' *Current Psychology* 37(3): 661–67. doi: 10.1007/s12144-016-9557-5.

Science and Technology Select Committee. 2019. 'Forensic Science and the Criminal Justice System: A Blueprint for Change'. Authority of the House of Lords. Accessed July 2019. https://publications.parliament.uk/pa/ld201719/ldselect/ldsctech/333/333.pdf.

Simons, Daniel J., and Christopher F. Chabris. 1999. 'Gorillas in Our Midst: Sustained Inattentional Blindness for Dynamic Events'. *Perception* 28(9): 1059–74. doi: 10.1068/p281059.

Smalarz, Laura, Stephanie Madon, Yueran Yang, Max Guyll, and Sarah Buck. 2016. 'The Perfect Match: Do Criminal Stereotypes Bias Forensic Evidence Analysis?' *Law and Human Behavior* 40(4): 420–29. doi: 10.1037/lhb0000190.

Troscianko, Tom, Alison Holmes, Jennifer Stillman, Majid Mirmehdi, Daniel Wright, and Anna Wilson. 2004. 'What Happens Next? The Predictability of Natural Behaviour Viewed Through CCTV Cameras'. *Perception* 33(1): 87–101. doi: 10.1068/p3402.

12

Probative value of gait analysis

Graham Jackson and Ivan Birch

The assessment of the probative value of items of evidence in a trial at court is primarily the role of the trier of fact, whether that be the jury, magistrate or judge as appropriate. However, when it comes to expert evidence, triers of fact are very unlikely to possess the necessary expert knowledge to assess competently and safely the weight to be attached to any expert evidence that is proffered at trial. Indeed, in the jurisdiction of England and Wales, the definition of an expert is that of one who provides information that the ordinary juror is unlikely to possess and which helps the trier of fact assess the significance of the expert evidence (*The Criminal Procedure Rules, The Criminal Practice Directions* 2015, Forensic Science Regulator 2019). However, until relatively recently, there was no widely accepted set of principles that an expert could use as a foundation for forming their expert opinions. There were certain legal requirements on experts, centred on the expert being independent and impartial, but nothing on how an expert should logically form an opinion from their observations, i.e. nothing on the process of sound inference.

Guidelines for the reporting of expert evaluative opinion were published in 2015 by the European Network of Forensic Science Institutes (ENFSI) (2015). These guidelines replicated to a large extent an earlier standard for the formulation of expert opinion published by the Association of Forensic Science Providers (AFSP) in 2009. The guidelines set out a procedure to help experts to assess logically, and to communicate, the potential probative value of their observations. The inferential paradigm underpinning these guidelines is probabilistic in nature, with the evaluation of a likelihood ratio being the key step in the process. While there has been significant penetration and uptake of the guidelines in many main-stream areas of forensic expert evidence, there has been limited application in specialist fields such as forensic gait analysis. A primer for courts on forensic gait analysis, published in 2017 (the Royal Society and the Royal Society of Edinburgh 2017), mentions briefly the AFSP standard for reporting opinions and gives an illustration of how a likelihood ratio may be derived. However, the illustration does not reflect accurately the process of logical evaluation of a likelihood ratio and may therefore give a false impression of that process. Furthermore, the overall philosophy of gait analysis is described in the primer as one of 'individualisation', a philosophy that has limitations and is prone to logical errors (Saks and Koehler 2005, Cole 2014, Thompson et al. 2018), and one which is not condoned by those working in forensic gait analysis.

AFSP AND ENFSI INFERENTIAL PARADIGM

A probabilistic approach to dealing with the interpretation and reporting of expert findings, as described in the ENFSI and AFSP guidelines, is not new in the forensic literature, having been extensively described and discussed over at least two decades. Numerous references are available; we cite two relatively recent examples (Jackson, Aitken, and Roberts 2015, Robertson, Vignaux, and Berger 2016), but there are many more.

In order to help readers focus on practical aspects of interpretation, the underlying theorem involved in a probabilistic approach and the associated formulae and notation to describe the concepts will not be reiterated – these are readily available in the references. While notation and formulae can help clarify the underlying argument, the aims in this section are (1) to describe in words alone the process of logical interpretation and (2) to demonstrate, through the use of a hypothetical case example, how this approach may be applied to gait analysis.

We do not assert that there is yet a clear model for handling gait analysis data to inform probabilities, but we do assert that principles of logical interpretation can still be applied to aid balanced evaluation of observations.

In broad terms, two generic types of expert opinion have been identified (Jackson et al. 2006, Jackson, Aitken, and Roberts 2015). The first generic form is called investigative opinion and is usually, but not exclusively, appropriate during the initial, investigative phase of an enquiry, usually prior to a suspect having been identified or arrested. The role of the expert when operating in this investigative mode is that of providing information to help the investigator clarify what happened at a scene of crime and to help identify suspects. In gait analysis, a typical investigative question would be of the type – "What can you tell me about the gait of the person seen in the crime scene CCTV footage?"

The second generic type of expert opinion is called evaluative opinion and is appropriate in those situations where a suspect has been apprehended or charged. The issues addressed by evaluative opinion are typically and generally:

"Has the person of interest carried out the particular activity of interest?" or
"Does this trace/recovered item come from the person of interest (or a specified item from that person)?"

Note that these questions, as phrased, relate to an issue involving a person of interest, whether that be a defendant, a suspect or a complainant. They are typically the sort of questions that are of importance to triers of fact in a court of law when they are fulfilling their duty of arriving at a decision on the ultimate issue of whether the defendant is guilty of the offence as charged. Following ENFSI and AFSP guidelines, the type of expert opinion given to help address this generic group of questions is an expression of the weight, or probative value, of the expert evidence in favour of one or other of the competing allegations. It is the expert's role to assess, i.e. evaluate, the probative value of the observations and to communicate that evaluation to help the trier of fact arrive at a decision. The trier of fact is perfectly entitled to accept, modify or reject the expert's evaluation.

In gait analysis, a typical evaluative issue would be of the form – "Is the person portrayed in the crime scene CCTV footage the defendant?"

This division into two generic groups of issue, and the two forms of opinion (investigative and evaluative) given to address the issues, arose from a realisation that experts may use different inferential processes when faced with different types of issue encountered within a case. This, in turn, implies two different roles for the expert – investigative and evaluative. However, it must be stressed that in real casework the two roles can, and do, intertwine; what is important is that the expert understands which type of issue is being addressed and, therefore, which type of inferential route and type of opinion is appropriate for the issue in question. Specifying the type of issue, and therefore the type of opinion, defines the knowledge and expertise required to form the opinion in a reliable and competent way. Reports and statements should explain for the client and the users of the expert evidence the rationale of the examination and the logical process by which the expert has arrived at an opinion.

While experts provide different types of opinion, depending on the type of issue under consideration, ENFSI and AFSP guidelines focus only on one type – evaluative opinion. The guidelines set out a formal procedure for forming such an opinion, but, probably because much less has been published on forming investigative opinion, ENFSI and AFSP guidelines provide very little advice in this area of work.

Investigative opinion can take various forms, including those of explanations for the expert observations and posterior probabilities for explanations. There are benefits and limitations of such forms of opinion of which the expert, and the receivers of expert opinion, should be aware. Perhaps the most serious limitation with both explanations and posterior probabilities is the real risk, unless mitigating procedures are applied, that cognitive bias will sway unfairly, unjustifiably and opaquely the opinion that is given by the expert.

It is not the intention here to explore investigative opinion in gait analysis further, but rather to focus on evaluative opinion, the type of issue most frequently encountered in court for gait analysts.

EVALUATIVE OPINION IN GAIT ANALYSIS

The basis of forming an evaluative opinion is the assessment of an entity called a likelihood ratio. The likelihood ratio is the ratio of two conditional probabilities, specified as follows:

1. the probability of the observations, given the truth of the prosecution proposition and dependent on (a) the relevant circumstances of the case and (b) the level of knowledge and expertise of the expert assigning the probability
2. the probability of the observations, given the truth of the defence proposition and dependent on (a) the relevant circumstances of the case and (b) the level of knowledge and expertise of the expert assigning the probability

Note the use of the word 'conditional'. The two probabilities are conditioned on, i.e. depend on, several factors, notably the truth of the propositions, the relevant circumstances and the expert's knowledge and expertise. The values assigned to these probabilities could well change if any of these factors change.

It follows from the definition of a likelihood ratio that, in order to provide such an evaluation, the expert needs to be provided with:

a) both a prosecution proposition and a defence alternative proposition,
b) details of the circumstances of the case that are relevant to, and which therefore condition, the expert's assignment of probabilities for the observations, and they need:

c) the knowledge and understanding of the evidence type, based on whatever reliable, relevant data there may be, in order to assign realistic, robust probabilities for the observations

A) Propositions

It is generally the case that the prosecution proposition will be known when a case is submitted to an expert. The police and/or prosecution will have made an allegation against a suspect or defendant, and the nature of the prosecution's contention will therefore be clear. On the other hand, the defence alternative proposition may not be available or may not have been communicated to the expert. In an ideal situation, the defence will have put forward an alternative proposition but, in the absence of such an alternative, and if the defendant is pleading 'not guilty', then the alternative proposition may be assumed to be "the person portrayed in the CCTV footage is not the defendant". In situations where a specific alternative proposition is put forward, then that alternative is the one that can be adopted by the expert in their evaluation of a likelihood ratio. If, for example, the defence assert that "the person portrayed in the CCTV footage is the brother of the defendant" or "the person portrayed in the CCTV footage is from the same 'gang' as the defendant", then those will be the alternative propositions adopted by the expert.

B) Relevant circumstances

Elements of the case circumstances that would have a bearing on the probabilities for the observations must be made known to the expert. In gait analysis, these elements would include factors such as distances, angles, lighting, recording equipment, recording parameters, the weather and the nature of the incident. All these 'conditioning' factors need to be taken into account by the expert when assigning probabilities for the observations, given the truth of the competing propositions.

C) Assignment of probabilities

There is considerable discussion in the academic world on the nature of probability and the extent to which the assignment of a value requires, and relies on, the availability of what may be called 'hard' data rather than the 'soft' data that reside in the memory of the expert. One school of thought (Biedermann et al. 2017) has the view that, because probability is a product of the mind, all probabilities are essentially subjective in nature, even if 'hard' data have been used. Under this view, the expert will use in an intelligent way whatever relevant data, whether 'hard' or 'soft', there may be, and accommodates through their own expert knowledge and understanding the limitations of the available data. The other school of thought (Morrison and Enzinger 2016) is of the view that probabilities should be assigned on the basis of 'hard', relevant, frequentist-type data. In the absence of such 'hard' data, the expert is unable to assign reliable probabilities. The different views are not yet reconciled, but what is beyond doubt is that the probabilities to be assigned relate to observations that are the result of a singular event – the case under consideration, perpetrated by an individual offender. It is not possible to re-run the case many times in order to obtain a frequency of occurrence of particular outcomes (observations). Inevitably, the expert has to deal with the uncertainties involved when assigning probabilities for the outcomes of a singular, non-reproducible event. The key issue is – if there are data, whether 'hard' or 'soft', how relevant and reliable are those data for the case in hand?

If the available data have been obtained from samples that mimic the conditions of the case in hand, then the probabilities assigned from those data will be more reliable than those obtained from samples whose conditions are further away from those of the case in hand. In the latter instance, the expert needs to use judgement, based on expert knowledge and understanding, to adjust the probabilities accordingly. Sensitivity analysis can be used to explore the impact of varying the probabilities on the magnitude of the resulting likelihood ratio.

In the absence of relevant 'hard' data, the expert can only rely on the 'soft' data provided by experience of similar situations to assign probabilities. The robustness of such probabilities depends on the expert having a sufficiently large body of relevant experience and on being able to mitigate the effects of poor cognitive performance and cognitive bias in memory and recall. We are not arguing that such 'soft' data should not be used but, rather, that caution should be exercised if they are to be used. Whatever way an expert arrives at their opinion, a valuable measure of the reliability of their opinion would achieve through calibration of that expert's opinions. To the best of the authors' knowledge, this was first implemented in the field of handwriting examination by Found and Rogers (2003) and has been recommended for all other fields of expert evidence (Evett et al. 2017).

EVALUATING GAIT OBSERVATIONS

The next section illustrates how the ENFSI/AFSP probabilistic inferential framework may be applied in cases that require an evaluative opinion in gait analysis. Our example is based on a hypothetical case. The first version of the case utilises relatively straightforward case circumstances and observations in order to illustrate the principles. The second version of the case explores the complications that arise in more realistic situations when the case circumstances and observations are not as straightforward.

Version 1
The circumstances of the case are as follows.

In a case of serious assault, CCTV footage was recovered from one camera on a street leading away from the scene of the attack. The footage was at an appropriate time to have recorded the perpetrator leaving the scene, and the police had strong reasons to believe that the sole figure depicted in the footage was the actual perpetrator. The footage of this figure will be designated as the questioned footage. Police also recovered CCTV footage from the same camera on each of four days after the assault. One of the many figures depicted in this footage is a person who the police suspect, based on other information, as being the person of interest in the questioned footage. The lighting conditions on all days were very similar. On initial inspection, it was believed there was sufficient quantity and quality of footage in both the questioned and reference footage to provide a reliable picture of the features of gait of the persons of interest.

Following ENFSI guidelines, the first stage of the evaluation process requires the provider to establish the fact in issue of the immediate client (in this case, the police), taking into account the needs of other stakeholders such as the court, the victim's family, the suspect and the wider community. In this particular case, the police alleged that the figure seen in the questioned footage was the suspect. The suspect denied the allegation and said that he had nothing whatsoever to do with the crime and was not the figure seen in the questioned footage. He agrees that the reference footage was of him. The fact in issue in this case can then be specified as – "Is the figure in the questioned footage the suspect?" However, the fact in issue in other cases may not be as straightforward, and would require more discussion and agreement between the parties on what is the actual issue in the case.

Having established the fact in issue, the next stage of the guidelines requires the provider to define the prosecution proposition and at least one alternative proposition based on the case circumstances and the defence assertions. In this case, the prosecution proposition seems straightforward – "The figure in the questioned footage is the suspect" – and, given the suspect's denial, the alternative proposition can be specified as – "The figure in the questioned footage is not the suspect".

The next stage of the guidelines requires the expert to assign probabilities for their observations, given the truth of the competing propositions.

Following recommended practice in gait analysis, the questioned footage was viewed before the reference footage and features of gait recorded. Ideally, the method of categorisation of observed features should be guided by a standard framework. The categorisation of features should be unambiguous and, ideally, designed to ensure that the outcomes (the actual observations made within each category of feature) are mutually exclusive and exhaustive. If these conditions are met, then population datasets can be compiled using the framework to help the estimation of frequencies of occurrence of outcomes. One published framework is a 'features of gait' tool specifically designed for use in observational gait analysis in a forensic context (Birch et al. 2019). The tool was developed based on research into the use of observational gait analysis in the clinical context and on the professional experience of forensic gait analysts. It facilitates a repeatable and reproducible approach for each piece of footage viewed. The tool can be used as a stand-alone item, or in conjunction with a database, as reported by Birch, Gwinnett, and Walker (2016). This database compiles the features observed in approximately 1000 individuals, of whom approximately 600 were male. Of the 600 males, approximately 500 were in the age range 18 to 50 years old. The size of the database, in terms of the number of individuals comprising it, is impressive. However, a limitation is that the features of gait have been recorded from observations of a single sequence of walking for each individual. Hence, knowledge of the degree of 'within sample' variation for each individual is limited.

For our case example, we will use the categories of features and specific outcomes as detailed in the tool. Assume the following outcomes were observed in the questioned footage:

- an asymmetrical gait
- shoulders relatively level
- right knee points slightly outwards, away from the midline of the body, at feet adjacent during swing
- left knee points slightly outwards, away from the midline of the body, at feet adjacent during swing
- a narrow base of gait
- right foot points outwards, away from the midline of the body, at feet adjacent during swing

The reference footage was then viewed and all the same features as in the questioned footage were observed consistently in the reference footage over the four days. No differences between the questioned and reference footage were observed.

In order to assign the two probabilities of the likelihood ratio, the expert needs to ask the following two questions:

1) "If the prosecution proposition were true, what is my probability for obtaining the observations from the questioned footage, given my knowledge of the reference footage, given the relevant case circumstances, and given my expert knowledge of gait?"
The value assigned for this probability forms the numerator of the likelihood ratio.

2) "If the defence proposition were true, what is my probability for obtaining these observations from the questioned footage, given the relevant case circumstances, and given my expert knowledge of gait?"
The value assigned for this probability forms the denominator of the likelihood ratio.

The guidelines recommend the use, wherever possible, of relevant, appropriate, published data to support assignment of probabilities, but the guidelines do not rule out the use of unpublished data, or of personal data based on experience, provided their use is justifiable and is made clear to the receivers of the opinion.

In considering an answer to the first question (for the numerator of the likelihood ratio), the expert needs to take into account the variability in the gait of the suspect. Are the observations from the questioned footage to be expected, given the questioned footage is actually of the suspect and given what has been seen in the reference footage of the suspect over the four days subsequent to the incident? The reference footage shows a high degree of consistency in the suspect's gait over the four days. The expert is therefore of the opinion that the questioned footage is just what they would expect if that footage were truly of the suspect. It is practically certain that the questioned footage would appear as it does, given it is of the suspect. They therefore assign a probability approaching a value of 1 (or 100%) for the numerator of the likelihood ratio. Had there been variation in the suspect's gait, then the expert may think a probability somewhat less than a value of 1 would be appropriate.

Turning now to an answer to the second question (for the denominator of the likelihood ratio), the expert's probability for obtaining the observations from the questioned footage will be informed by the frequency of occurrence of those observed features in a relevant population. The key consideration is – "What is the relevant population?" (Champod, Evett, and Jackson 2004). Under the condition of the alternative proposition – "The figure in the questioned footage is not the suspect" – it is implied that the figure is instead a member of a population that could be considered as potential perpetrators. This in turn begs the question of what sort of person could be considered as a potential perpetrator. Of course, if there are witnesses to the attack, the population would be informed by their evidence (provided that that evidence is accepted by the court). Depending on the precise circumstances of a case, certain sectors of society may be ruled out on grounds of, for example, age or gender. Once a relevant population has been defined, and assuming there are good reference data on that population, then a relative frequency of occurrence of the combination of features observed in the questioned footage could be derived from those data. In the absence of any dataset, the expert could resort to their experience and knowledge to try to recall the incidence of the observed features in a relevant population. The problem here lies with the extent and relevance of the experience of the expert, and with the ability of the expert to recall faithfully the incidence of features. Such personal assessment of frequencies of occurrence, and the probabilities that may be derived from them, could be described as being based on 'soft' data, i.e. those that exist in the memory of the expert, as opposed to those 'hard' data that exist in physical or digital databases that can be interrogated, shared, reproduced and tested.

Returning to our example, the expert could adopt this 'soft data' approach to estimate the number of times they had observed, among the total number of relevant people they had observed in their professional experience, the combination of features seen in the questioned footage. However, this process of estimation has potentially serious limitations, not least of which are the risks of cognitive bias and of inefficient recall of features. On the other hand, if they wished to use 'hard data', what datasets are available? The answer is – very few. Mention has been made earlier of one database (Birch, Gwinnett, and Walker 2016), which is aligned to the Sheffield Features of Gait Tool (Birch et al. 2019) (see Appendix 1). The expert would need to consider whether this dataset is of a relevant population and that the conditions of recording and categorising the features are similar to those used on the questioned footage. For the purposes of illustration, let us assume the court would accept that the perpetrator was male and of the age range 18–50. On interrogating the database, let us assume there were 10 individuals out of a total of 500 males between the ages of 18 and 50 who were recorded as having the same combination of features as seen in the questioned footage. The data entries for each of the 10 males would contain further features of gait, but those categories of features had not been observed in the questioned footage and are therefore irrelevant to the evaluation. The relative frequency of 10/500, or 0.02 (2%), for the observed combination of features then informs the expert's assignment of probability. The expert is of the view that the database does provide them with reliable, robust, relevant figures and they therefore assign a value of 0.02 (2%), based on that relative frequency, for the denominator probability.

Note that dependency between features has been accommodated by searching the database for the combination of features. However, if the frequencies of individual features were to be assessed individually and sequentially, then the frequency of occurrence of any one feature must be conditioned on the occurrence of the preceding feature (or combination of features). For example, the database would be interrogated firstly to assess how common was feature A and then, looking solely at those entries that showed feature A, the subset would be interrogated to assess how many also show feature B. In a further subset that showed both features A and B, the data would be interrogated for entries that also showed C. And so on, until all features had been assessed. This is an essential procedure in order to take account of the inevitable relationship between some features.

The expert now has assigned the two probabilities that form the likelihood ratio – the numerator is close to 1 and the denominator is 0.02. The ratio of these two values, the likelihood ratio, is approximately 50. While the data on which the denominator is based are what we are calling 'hard' data, the probability assigned to the denominator is nonetheless a subjective choice in which expert judgement and personal decision-making by the expert play large roles. The same is true of the numerator, particularly if no data about the reproducibility of the appearance of the suspect's gait are available. It is important to stress that it is not possible, generally, to assign a value with a high degree of precision for these probabilities and, hence, for the likelihood ratio. What is important is the order of magnitude of the likelihood ratio. In this case, the value of 50 for the likelihood ratio falls in the ENFSI/AFSP category (European Network of Forensic Science Institutes 2015; Association of Forensic Science Providers 2009) of "moderate support" for the proposition that the figure in the questioned footage is the suspect rather than some other, unknown, male person (between the ages of 18–50).

As an aside, and as an illustration of alternative sources of data to inform the denominator, the expert might consider another, potentially very relevant database, viz. the footage of people, including the suspect, recorded by the CCTV cameras during the days after the incident. On examination of the rest of this footage, there were approximately 200 males, roughly of the age range 18–50, for whom features of gait could be assessed reliably. On compiling the observations, two people, including the suspect, were observed to have the same combination of features as observed in the questioned footage. The relative frequency of the combination of features in this database is therefore 2/200 or 0.01 (1%). While this figure would provide a smaller probability for the denominator (0.01), and therefore a larger likelihood ratio against the defendant (100), it is based on a smaller dataset than that used in the original evaluation. It may be reasonable for the expert to act conservatively (in favour of the defendant) and to stick with a value of 0.02 for the denominator probability that is based on a larger database.

The expert should describe in their report their rationale for assigning the values for the probabilities, and therefore the likelihood ratio, and give a conclusion that relates back to the fact in issue and the propositions as described at the beginning of the report. Such a conclusion could be phrased along the lines of:

"The observations made on the questioned footage are in my opinion approximately 50 times more probable if the questioned footage is of the suspect rather than of some other, unknown, male person in age range 18–50 years old."

If a verbal equivalent for the likelihood ratio is to be reported, then typical phraseology would be:

"In my opinion, the observations provide moderate support for the proposition that the questioned footage is of the suspect rather than some other, unknown, male person in age range 18–50 years old."

Version 2

Real-life casework is rarely as straightforward as depicted in Version 1. If we change the case circumstances and the observations, we can explore the implications for a case nearer to reality.

In a case of serious assault, CCTV footage was recovered from one camera on a street leading away from the scene of the attack. The footage was at an appropriate time to have recorded the perpetrator leaving the scene, and the police had strong reasons to believe that the sole figure shown in the footage was the actual perpetrator. The footage of this figure will be designated as the questioned footage. Based on other information, police arrest a man who they believe is the perpetrator. They recover footage from several security cameras on premises near to where the man lives. These cameras have recorded the man walking to and from his house over the course of the previous week or so. As in Version 1, the suspect denied the allegation and said that he had nothing whatsoever to do with the crime and was not the figure seen in the questioned footage. He does agree that the reference footage was of him. The fact in issue in this version remains the same as in Version 1 – "Is the figure in the questioned footage the suspect?" – and the pair of propositions to be addressed is also the same:

"The figure in the questioned footage is the suspect" and
"The figure in the questioned footage is not the suspect".

On viewing the relatively short length of questioned footage, the following features of gait were observed and recorded:

- an asymmetrical gait
- shoulders relatively level
- right knee points slightly outwards, away from the midline of the body, at feet adjacent during swing
- left knee points slightly outwards, away from the midline of the body, at feet adjacent during swing
- a narrow base of gait
- right foot points outwards, away from the midline of the body, at feet adjacent during swing

These features were seen consistently throughout the full sequence of steps in the questioned footage.

The reference footage comprised 10 sequences from the five days, thereby providing many more steps than in the questioned footage. Some variation in the occurrence of features between different sections of the reference footage was observed. For example, the symmetry of gait was recorded as 'asymmetrical' in 90% of the sequences and 'symmetrical' in 10%. In comparison, the questioned footage was recorded simply as 'asymmetrical'.

The relative complexity of the observations in the reference footage in this version of the case requires a more sophisticated approach to interpretation than in Version 1. While the proposition pair remains the same as in Version 1, the assignment of conditional probabilities is not as straightforward. In the first version, the observations (the outcomes) were categorical, showing no variation either within the questioned and the reference footage or between the two sets of footage. The observations made in this second version of the case did show variation within the reference footage.

Such variation, potentially, in both questioned and reference footage, can be accommodated probabilistically. However, at this stage of development of interpretation of gait, no probabilistic model of the required level of sophistication is available. We are confident that such a model is feasible and will be developed.

Despite the lack of such a model, some progress can still be made in our case example because, as deliberately designed, it is only the reference footage, not the questioned footage, that shows variation in features. Consideration of the reference footage comes into play only when considering the numerator – the numerator probability is conditioned on the information in the reference footage; the denominator is not influenced at all by the reference footage. Therefore, we can follow the procedure as in Version 1 for assigning a probability for the denominator, but we will describe a different approach to assigning a probability for the numerator.

Firstly, the variation within the 10 sequences in the reference footage would need to be defined. To this end, the percentage occurrence of the different outcomes within each observed category of feature were compiled (Table 12.1).

For the numerator, the question for the expert is – "What is my probability for obtaining the observations from the questioned footage, if the figure in that footage is truly the suspect, given my knowledge of the reference footage of the suspect's gait, given the relevant case circumstances, and given my expert knowledge of gait?" This question can be broken down into sub-questions for each category of feature. Looking at the first category, that of 'symmetry', and the outcome 'asymmetrical' that was observed, the sub-question would be – "What is my probability of observing asymmetry

TABLE 12.1
Percentage Frequency of Occurrence of Observations of Features of Gait in 10 Sequences of Reference Footage

Feature	Percentage occurrence for each outcome (N=10)	
Symmetry of gait	Asymmetrical	Symmetrical
	90	10
Comparative height of shoulders	Level	Left lower than right
	80	20
Orientation of right knee at feet adjacent during swing	Outward	Inward
	100	0
Orientation of left knee at feet adjacent during swing	Outward	Inward
	80	20
Base of gait	Narrow	Moderate
	20	80
Orientation of right foot at feet adjacent during swing	Outward	Inward
	60	40

in the questioned footage if that footage is truly that of the suspect, given my knowledge of the reference footage of the suspect's gait (and given the relevant case circumstances)?" The data from the reference footage indicated the suspect exhibited asymmetrical gait on 90% of the sequences. In the absence of any other information on the suspect's gait, it would then seem reasonable to assign a probability of 0.9 for observing asymmetry in the questioned footage if that footage truly is of him. Taking the second category, that of 'comparative height of shoulders' and the outcome 'relatively level', the sub-question would be – "What is my probability of observing relatively level shoulders in the questioned footage if that footage is truly that of the suspect, given my knowledge of the reference footage of the suspect's gait (and given the relevant case circumstances)?" Following similar logic to the first sub-question, it may seem that observing level shoulders in the questioned footage could be assigned a probability of 0.8, given the data in Table 12.1 for this feature. However, in order to account for any dependency (as mentioned for Version 1 of the case), this probability needs to be conditioned on the previous observation of asymmetry. The sub-question should be phrased along the lines of – "What is my probability of observing relatively level shoulders in the questioned footage if that footage is truly that of the suspect, given that I have observed asymmetry in gait and given my knowledge of the reference footage of the suspect's gait (and given the relevant case circumstances)?" The existence and degree of dependency between the two features (strictly between the 'outcomes') is a judgement for the expert based on their knowledge, experience and whatever relevant data there may be. The extent and impact of dependency may be best understood through the testing of a sufficient quantity of observational data, ideally of the suspect himself. However, in the absence of such data on the suspect, then data from a population relevant to the suspect could provide informative statistics.

For the purposes of this example, and purely to keep the illustration uncomplicated, it will be assumed there is complete independence between all variables (a situation that patently is not true). If we assume complete independence, then the probabilities for each outcome may be multiplied together to give an overall probability for observing the combination of features seen in the questioned footage. Using the data in Table 12.1, this probability will be the product of the individual probabilities 0.9, 0.8, 1, 0.8, 0.2, 0.6. The numerator probability can therefore be assigned a value of 0.07.

Turning to the denominator, the expert asks – "If the defence proposition were true, what is my probability for obtaining these observations in the questioned footage, given the relevant case circumstances and given my expert knowledge of gait?" Note there is no conditioning on the information in the reference footage because, given this alternative proposition, the suspect has nothing at all to do with the incident. On the defence alternative, the figure in the questioned footage is not the suspect, it is someone else (of a relevant population). The assignment of a probability follows a similar procedure as in Version 1 and, given we have the same observations from the questioned footage as Version 1, a value of 0.02 may be assigned based on the frequency of occurrence of the combination of features.

The likelihood ratio is therefore 0.07 divided by 0.02, i.e. 3.5. In words, a likelihood ratio of 3.5 means that the observations are approximately three and a half times more probable if the prosecution proposition, rather than the defence proposition, were true. And in terms of a verbal scale of support (European Network of Forensic Science Institutes 2015), the likelihood ratio can be expressed as offering 'limited' support in favour of the prosecution proposition rather than the defence proposition.

We believe the lack of database does not preclude an expert from giving an evaluative opinion. An expert is entitled to offer an expert opinion, provided it is made clear to the receivers that it is not based on any, to paraphrase, 'hard' data, but is based on the expert's experience and judgement. What is important is that:

1) the opinion is based on consideration of a pair of propositions (to demonstrate balance and impartiality)
2) the opinion is conditional on the relevant case circumstances
3) the expert has assigned broad probabilities for the observations under the two competing propositions and has expressed verbally the ratio of the two probabilities

In such circumstances, an opinion may be expressed along the following lines:

"The observations made on the questioned footage are, in my opinion, probable if the questioned footage is of the suspect rather than of some other, unknown, male person in age range 18–50 years old.

In that respect, the observations provide support for the proposition that the questioned footage is of the suspect rather than some other, unknown, male person in age range 18–50 years old. However, I am unable to quantify the degree of support."

Of course, the experience, expertise and past performance of the expert may be explored and tested by the court, if those aspects have not already been declared, before the court accepts the validity and reliability of the opinion.

SUMMARY

There is undoubtedly a complexity and interdependence of factors that determine the appearance of a person's gait. Sophisticated statistical models for assigning conditional probabilities of observations to assist in the evaluation of gait evidence are required. However, even in the absence of such models, there should be no barrier to adopting a likelihood ratio approach to evaluating gait evidence.

We have shown how a likelihood ratio may be evaluated using appropriate pairs of case-specific propositions and expert-assigned conditional probabilities. At one end of the spectrum of expert knowledge, probabilities are based on the accumulated experience of the expert, and, at the other end of the spectrum, they are based on expert use of data that are relevant to the propositions. Probabilities of the former type are inevitably imprecise and more prone to cognitive bias than the latter type.

REFERENCES

Association of Forensic Science Providers. 2009. "Standards for the formulation of evaluative forensic science expert opinion." *Science and Justice* 49(3):161–164.

Biedermann, Alex, Silvia Bozza, Franco Taroni, and Colin Aitken. 2017. "The meaning of justified subjectivism and its role in the reconciliation of recent disagreements over forensic probabilism." *Science and Justice* 57(6):477–483.

Birch, Ivan, Claire Gwinnett, and Jeremy Walker. 2016. "Aiding the interpretation of forensic gait analysis: Development of a features of gait database." *Science and Justice* 56(6):426–430.

Birch, Ivan, Maria Birch, Lucy Rutler, Sarah Brown, Libertad Rodriguez Burgos, Bert Otten, and Mickey Wiedemeijer. 2019. "The repeatability and reproducibility of the Sheffield Features of Gait Tool." *Science and Justice* 59(5):544–551.

Champod, Christophe, Ian W. Evett, and Graham Jackson. 2004. "Establishing the most appropriate databases for addressing source level propositions." *Science and Justice* 44(3):153–164.

Cole, Simon A. 2014. "Individualization is dead, long live individualization! Reforms of reporting practices for fingerprint analysis in the United States." *Law, Probability and Risk* 13(2):117–150.

The Criminal Procedure Rules. The Criminal Practice Directions. October 2015 edition. Amended April, October and November 2016; February, April, August, October and November 2017; April and October 2018; and April 2019. Accessed July 2019. www.justice.gov.uk/courts/procedure-rules/criminal/rulesmenu-2015

European Network of Forensic Science Institutes. 2015. "ENFSI guideline for evaluative reporting in forensic science, strengthening the evaluation of forensic results across Europe (STEOFRAE)." Dublin. Accessed July 2019. http://enfsi.eu/wp-content/uploads/2016/09/m1_guideline.pdf

Evett, Ian W., C. E. H. Berger, J. S. Buckleton, C. Champod, and Graham Jackson. 2017. "Finding the way forward for forensic science in the US—A commentary on the PCAST report." *Forensic Science International* 278:16–23.

Forensic Science Regulator. 2019. "Legal obligations FSR-I-400." Issue 7. Birmingham: Forensic Science Regulator.

Found, Bryan James, and Douglas Kelman Rogers. 2003. "The initial profiling trial of a program to characterize forensic handwriting examiners' skill." Long Beach, CA: American Society of Questioned Document Examiners.

Jackson, Graham, Stella Jones, Gareth Booth, Christophe Champod, and Ian W. Evett. 2006. "The nature of forensic science opinion – A possible framework to guide thinking and practice in investigations and in court proceedings." *Science and Justice* 46(1):33–44.

Jackson, Graham, Colin Aitken, and Paul Roberts. 2015. "Case assessment and interpretation of expert evidence: Guidance for judges, lawyers, forensic scientists and expert witnesses." *Practitioner Guide* (4). London: Royal Statistical Society's Working Group on Statistics and the Law.

Morrison, Geoffrey Stewart, and Ewald Enzinger. 2016. "What should a forensic practitioner's likelihood ratio be?" *Science and Justice* 56(5):374–379.

Robertson, Bernard, George A. Vignaux, and Charles E. H. Berger. 2016. *Interpreting Evidence: Evaluating Forensic Science in the Courtroom.* Chichester: John Wiley & Sons.

Saks, Michael J., and Jonathan J. Koehler. 2005. "The coming paradigm shift in forensic identification science." *Science* 309(5736):892–895.

The Royal Society and the Royal Society of Edinburgh. 2017. *Forensic Gait Analysis: A Primer for Courts.* London: The Royal Society.

Thompson, William C., Joelle Vuille, Franco Taroni, and Alex Biedermann. 2018. "After uniqueness: The evolution of forensic science opinions." *Judicature* 102(1):18–27.

13

Case studies

CASE STUDY 1: BERRY'S ONE STOP STORE ROBBERY

Michael Nirenberg

Background: One evening three armed individuals robbed Berry's One Stop Store in Wayne County, Tennessee. The perpetrators were dressed in black and wore handkerchiefs over their faces to hide their identities. One of the investigators reviewing surveillance video of the robbery recognised that one of the thieves had what they considered to be a distinctive gait. They noticed that the feet of one of the perpetrators were excessively out-toed, that is to say pointing outwards away from the mid line of the body when weight bearing (see Figure 13.1).

The investigator then reviewed surveillance video from the store taken the day before the crime and saw a customer, that he knew, who appeared to walk in a manner similar to the perpetrator who had an out-toed gait.

Question: The forensic gait analyst was tasked with offering an opinion as to whether the known person, the customer in the store the day prior to the robbery, walked in the same way as the perpetrator with the distinctive gait, and could therefore be the same person.

190 Chapter 13. Case studies

FIGURE 13.1 The perpetrator demonstrating an out-toed position of the feet, in this instance when standing.

Materials: The analyst was provided with the questioned footage of the robbery, which comprised of footage from 10 cameras showing various views of the interior and exterior of the store, and reference footage of the suspect. This case was somewhat atypical as the video of the perpetrator and the video of the person of interest were taken from the same cameras, thus reducing limitations that might typically occur from variations in camera direction, angle, lens distortion, and other distinct features of the video equipment.

Analysis and comparison: The forensic gait analyst initially performed a preliminary review of the questioned footage and determined that, although some of the video was not suitable for use in forensic gait analysis, sufficient footage was suitable to make a comparison. The unsuitable footage was rejected for a number of reasons including the image being too dark to see clearly the perpetrator's gait (outdoor camera), the perpetrator running (without footage of the customer running for comparison), and/or the perpetrator/suspect not being shown taking an adequate number of mid steps to establish their normal gait.

For each usable segment of footage of the robbery, the expert utilised the Sheffield Features of Gait Tool (see Chapter 7 and Appendix 1), which provides a systematic, validated method of analysing a person's gait. The footage of the perpetrator, the questioned footage, was analysed first, in order to reduce cognitive bias (see Chapter 11). For each camera with suitable questioned footage, the Sheffield tool was used to guide the observation of the perpetrator's gait and note the features of gait seen. The analyst worked through all the footage captured by a single surveillance camera, using the tool and noting each feature of gait from head to foot. The analyst then moved to the next camera's footage, applying the same process until all suitable footage was analysed.

Once all of the questioned footage had been analysed, the analyst took a few days' break from his work on the case to allow details of the robber's gait to fade from memory and reduce the possibility of cognitive bias. The expert then used the Sheffield Features of Gait Tool to analyse all of the reference footage. During the process of analysing the reference footage, the expert did not refer back to the questioned footage, again reducing the risk of cognitive bias.

With the analysis of both the questioned and reference footage complete, the analyst then summarised their observations for the figure in the questioned footage, and the subject in the reference footage. He then compared these summarised observations (see Figure 13.2). The expert concluded that the figure in the questioned footage and the subject in the reference footage had no features of gait that precluded them from being the same person, and that their gait was similar enough for the forensic gait analysis evidence to provide limited support for the proposition that they were the same person. This conclusion was based on the use of the verbal expressions of probative value of support or rejection (see Chapter 8).

The conclusion took into account the limitations of the footage, which included:

a. the limited number of consecutive steps shown in the footage
b. the number of tables, counters, chairs, and other obstacles in the store around which the robber had to navigate and which may have affected the figure/subject's gait (Figure 13.1)
c. the figure in the questioned footage was carrying an object, which may have affected their gait
d. the clothing worn by the figure/subject in the footage appeared relatively loose fitting, which limited the observation of some features of gait
e. the limited resolution and frame rate of the footage

The analyst's report was submitted for independent verification prior to submission to the law enforcement agency that had commissioned the work.

Outcome: The law enforcement agency forwarded the case information and the forensic gait analyst's report to the district attorney, who obtained an arrest warrant for the suspect. When the suspect was arrested and presented with the gait evidence (and cell phone-related data), he confessed and revealed the identities of the other two robbers (Brentzel, 2018).

Learning points:

1. Gait analysis is not standing analysis. During the robbery, the perpetrator in question stands with his feet pointed significantly outward. The gait analyst should be careful not to misinterpret

192 Chapter 13. Case studies

	feature observed							Figure in the Questioned Footage	Subject in the Reference Footage
		✓		m oss R	marked on some steps right side only		sl mos L	slight on the majority of steps left side only	

		feature observed	Figure in the Questioned Footage	Subject in the Reference Footage
1	A1	symmetrical gait[1]		✓
	A2	asymmetrical gait[2]		
	A3	erratic gait		
		symmetry of gait could not be determined		
2	B1	no significant rolling of the head and torso[3]	✓	
	B2	rolling of the head and torso	✓	✓
	B3	rolling of the head and torso, more to the right on right steps than to the left on left steps		
	B4	rolling of the head and torso, more to the left on left steps than to the right on right steps		
		frontal plane motion of the head and torso could not be determined[4]		
3	B5	no significant yawing of the torso[5]		
	B6	yawing of the torso, right side forwards on left steps, left side forwards on right steps	✓	
	B7	yawing of the torso, right side forwards on right steps, left side forwards on left steps		
		transverse plane motion of the torso could not be determined[6]		
	B8	vertical movement of the head and torso on each step		✓

[1] A symmetrical gait is one in which the gross movement of the head and torso to the right is approximately the same as the gross movement to the left.
[2] An asymmetrical gait is one in which the gross movement of the head and torso to the right is different to the gross movement to the left.
[3] Rolling of the head and torso is a rotation that would be best seen from the front or back.
[4] The frontal plane is the plane of body movements which are best seen from the front or rear of the body.
[5] Yawing of the torso is a rotation that would be best seen from above.
[6] The transverse plane is the plane of body movements which are best seen from above.

FIGURE 13.2 Features of gait observed for the perpetrator (figure in the questioned footage) and the suspect (subject in the reference footage).

4	C1	head held approximately in line with the midline of the torso in the frontal plane			✓
	C2	head held tilted to the right relative to the midline of the torso in the frontal plane			
	C3	head held tilted to the left relative to the midline of the torso in the frontal plane		✓	
		frontal plane alignment of the head could not be determined			
5	C4	head held approximately in line with the midline of the torso in the sagittal plane[7]		✓	
	C5	head held forwards of the midline of the torso in the sagittal plane			✓
	C6	head held tilted forwards relative to the midline of the torso in the sagittal plane			
	C7	head held tilted backwards relative to the midline of the torso in the sagittal plane			
		sagittal plane alignment of the head could not be determined			
6	D1	shoulders relatively level		✓	✓
	D2	dipping of the right shoulder on right steps more than the left shoulder on left steps			
	D3	dipping of the left shoulder on left steps more than the right shoulder on right steps			
	D4	right shoulder lower than the left shoulder			
	D5	left shoulder lower than the right shoulder			
		relative height of the shoulders could not be determined			
7	E1	full arm swing		✓	✓
	E2	limited arm swing			
	E3	erratic arm swing			
	E4	oblique path of motion of the hand, the hand more medial anteriorly than posteriorly			
	E5	more arm swing movement anterior of the torso than posterior of the torso			
	E6	arm swing movement anterior and posterior of the torso approximately equal			
	E7	elbows remain flexed throughout arm swing			
	E8	flexion at the elbow increasing during the forward motion of the arm			
		no right arm swings are visible			
		no left arm swings are visible			
		a reliable assessment of arm swing could not be made			

[7] The sagittal plane is the plane of body movements which are best seen from either side of the body.

FIGURE 13.2 Continued.

194 Chapter 13. Case studies

8	F1	largely linear motion occurring at right hip[8]	✓oss	✓oss
	F2	largely linear motion occurring at left hip		✓oss
	F3	largely linear motion occurring at the right hip, with slight circumduction occurring prior to heel strike[9]	✓oss	✓oss
	F4	largely linear motion occurring at the left hip, with slight circumduction occurring prior to heel strike	✓oss	✓oss
	F5	circumduction occurring at right hip		
	F6	circumduction occurring at left hip		
		nature of the movement occurring at the left hip could not be determined		
9	G1	right knee points inwards, towards the midline of the body, at feet adjacent during swing		
	G2	left knee points inwards, towards the midline of the body, at feet adjacent during swing		
	G3	right knee points forwards approximately in line with the direction of travel at feet adjacent during swing		
	G4	left knee points forwards approximately in line with the direction of travel at feet adjacent during swing		
	G5	right knee points forwards, or slightly outwards, at feet adjacent during swing	✓	✓
	G6	left knee points forwards, or slightly outwards, at feet adjacent during swing	✓	✓
	G7	right knee points outwards, away from the midline of the body, at feet adjacent during swing		
	G8	left knee points outwards, away from the midline of the body, at feet adjacent during swing		
	G9	right more so than left		
	G10	left more so than right		
		orientation of the left knee at feet adjacent during the swing phase of gait could not be determined		
10	H1	right knee is flexed at or close to heel strike		
	H2	left knee is flexed at or close to heel strike		
	H3	right knee reaches a position of full or close to full extension at or close to heel strike	✓	✓
	H4	left knee reaches a position of full or close to full extension at or close to heel strike	✓	✓
	H5	right knee reaches a position of full or hyper extension at or close to heel strike		
	H6	left knee reaches a position of full or hyper extension at or close to heel strike[10]		
		angulation of the knees at or close to heel strike could not be determined		

[8] Linear movement in this instance is a movement occurring at the hip joint that results in the lower limb moving forward in an approximately straight line during the swing phase of gait.
[9] Circumduction in this instance is a movement occurring at the hip joint that results in the lower limb moving forward in an arc during the swing phase of gait.
[10] Full extension of the knee is a position in which the thigh and shank are aligned. Movement beyond this point is known as hyperextension, the knee appearing to fold backwards.

FIGURE 13.2 Continued.

11	I1	right knee is flexed prior to heel rise		
	I2	left knee is flexed prior to heel rise		
	I3	right knee reaches a position of full or close to full extension prior to heel rise	✓	
	I4	left knee reaches a position of full or close to full extension prior to heel rise	✓	✓
	I5	right knee reaches a position of full or hyper extension prior to heel rise		
	I6	left knee reaches a position of full or hyper extension prior to heel rise		
		angulation of the knees prior to heel rise could not be determined		
12	J1	base of gait is very narrow[11]	✓	✓
	J2	base of gait is narrow		
	J3	base of gait is moderate		
	J4	base of gait is wide		
	J5	base of gait is variable		
		base of gait could not be determined		
13	K1	right foot points inwards, towards the midline of the body, when weight bearing during stance		
	K2	left foot points inwards, towards the midline of the body, when weight bearing during stance		
	K3	right foot points forwards, approximately in line with the direction of travel, when weight bearing during stance		
	K4	left foot points forwards, approximately in line with the direction of travel, when weight bearing during stance		
	K5	right foot points forwards, or slightly inwards/outwards, when weight bearing during stance		
	K6	left foot points forwards, or slightly outwards, when weight bearing during stance		
	K7	right foot points outwards, away from the midline of the body, when weight bearing during stance	✓m	✓m
	K8	left foot points outwards, away from the midline of the body, when weight bearing during stance	✓m	✓m
	K9	right more so than left		
	K10	left more so than right		
		the comparative orientation of the feet, when weight bearing during stance, could not be determined		
		orientation of the feet when, weight bearing during stance, could not be determined		

[11] The base of gait is the side to side distance between the heels on consecutive steps.

FIGURE 13.2 Continued.

14			
	L1	right foot points inwards, towards the midline of the body, at feet adjacent during swing	
	L2	left foot points inwards, towards the midline of the body, at feet adjacent during swing	
	L3	right foot points forwards, approximately in line with the direction of travel, at feet adjacent during swing	
	L4	left foot points forwards, approximately in line with the direction of travel, at feet adjacent during swing	
	L5	right foot points forwards, or slightly inwards/outwards, at feet adjacent during swing	
	L6	left foot points forwards, or slightly outwards, at feet adjacent during swing	
	L7	right foot points outwards, away from the midline of the body, at feet adjacent during swing	✓m
	L8	left foot points outwards, away from the midline of the body, at feet adjacent during swing	✓m
	L9	right more so than left	
	L10	left more so than right	
		the comparative orientation of the feet, at feet adjacent during swing, could not be determined	
		orientation of the feet, at feet adjacent during swing, could not be determined	
	M1	right foot is noticeably inverted prior to heel strike[12]	
	M2	left foot is noticeably inverted prior to heel strike	
	M3	right foot noticeably everts on weight bearing	
	M4	left foot noticeably everts on weight bearing	
	N1	right forefoot is markedly raised prior to heel strike	
	N2	left forefoot is markedly raised prior to heel strike	
	N3	right heel raise is relatively early	
	N4	left heel raise is relatively early	
	O1	right foot is noticeably abducted prior to heel strike[13]	✓
	O2	left foot is noticeably abducted prior to heel strike	✓
	P1	during the early part of the stance phase of gait, the thigh of both lower limbs appears slightly inverted[14]	

[12] An inverted foot describes an orientation of the foot in which the sole is turned towards the midline of the body when the foot is viewed from the front or back.
[13] An abducted foot describes an orientation of the foot in which the toes are further away from the midline of the body than the heel when the foot is viewed from above or below.
[14] An inverted thigh describes an orientation of the thigh in which the lower end of the thigh is closer to the midline of the body than the upper end of the thigh.

FIGURE 13.2 Continued.

the positions of body segments during standing as an indication of their position during gait. Standing and walking are two different mechanical processes and should not be compared. If a person stands at a bar with their legs crossed, would you expect them to walk with their legs crossed?
2. When using the Sheffield Features of Gait Tool, always work through one camera's footage entirely before moving on to other footage. When multiple pieces of footage are available from different cameras, it is not appropriate to examine one feature of gait from the Sheffield tool for the subject on each camera sequentially, and then move to the next gait feature. This increases the possibility of cognitive bias, by facilitating anticipation of what will be seen before the footage is analysed. Each camera's footage should be analysed completely using the tool from head to foot, before moving on to the footage from the next camera. This reduces the risk of cognitive bias by separating the occasions on which a particular feature of gait is investigated, interspersing the investigation with that of other features of gait.
3. Consider all camera footage as a whole. Although each piece or section of surveillance footage may show a limited number of usual mid gait steps, or allow the observation of a limited number of features of gait, the summation of the information gained from all the footage may allow the determination of a person's typical gait.
4. Gait analysis should only make like for like comparisons. As noted previously, the analyst did not use the footage of the perpetrator running because there was no footage showing the suspect running. Walking and running are two completely different locomotor activities, using different mechanical processes and movements, producing different forces and pressures, and placing different anatomical and physiological demands on the body. They should not therefore be compared. In this case walking was only compared to walking.
5. Use caution with all descriptive language used. The expert should use clear, objective language in their report when describing the figure in the questioned and the subject in the reference footage. For example, terms such as "he" or "she" should be avoided as they imply identification of the sex of the person. While the gait analyst may be certain of the sex of the person in the footage, they are not an expert in sex determination and should therefore avoid making an implicit judgment by the language they use. Careful consideration should also be given to any description of the clothing worn by the person in the footage. A gait analyst is not a clothing expert, and even the apparent colour of clothing and footwear can vary significantly in different light sources and when captured by different cameras.

CASE STUDY 2: ARSON OF A BUSINESS PREMISES
Maria Birch

Background: A perpetrator was captured on CCTV footage entering an almost empty warehouse. Once inside they disarmed the alarm, and then exited by a side door which they left unlocked. Subsequently, a second perpetrator was also captured on CCTV footage entering the building and setting it on fire. The police retrieved several pieces of footage from various cameras around the area. All the footage was time stamped, allowing the two perpetrators to be reliably tracked between each piece of footage. The police had four suspects in custody.

Question: The forensic gait analyst was asked to determine the possibility that either of the two perpetrators in the questioned footage were one of the four suspects in custody. This question was a complex one involving comparison of questioned footage showing perpetrator 1 with the reference footage showing each of the four suspects, and a comparison of the questioned footage showing perpetrator 2 again with the reference footage showing each of the four suspects. This was an eight-way comparison. None of the four suspects offered relevant information at interview.

Materials: The gait analyst was provided with CCTV footage taken from a number of cameras at, and in the vicinity of, the warehouse and reference footage captured by various cameras showing the suspects in the charge area and corridors of a custody suite. An appropriately experienced colleague of the forensic gait analyst undertook a preliminary assessment of all the footage the police had submitted. For the questioned footage, the resolution was good, the lighting varied from footage to footage, some being good and some poor, and the frame rate varied from good to adequate (12 to 9 frames per second). In the custody suite reference footage, the resolution varied from good to adequate, the lighting varied from very good to adequate, but the frame rate was poor for the purposes of forensic gait analysis (6 frames per second). These technical characteristics of the footage were considered likely to have a negative effect on the final outcome of the analysis, and would therefore need to be taken into consideration when conclusions were drawn.

It was established that there was enough footage suitable for use in forensic gait analysis for the analysis to be undertaken. The police provided information and still images which allowed the two perpetrators to be distinguished, and the four suspects to be identified.

Analysis and comparison: The Sheffield Gait Tool was used to assist in the systematic analysis of the gait of the first perpetrator in the questioned footage, followed by the analysis of the gait of the second perpetrator. Once all of the questioned footage had been considered, the gait of each of the four suspects in the reference footage was analysed in the same way. As has been detailed earlier in this book, all precautions were taken to reduce cognitive bias. Once all the footage had been analysed, and the features of gait noted, the features of gait from the various pieces of questioned footage were summarised, producing a list of those features that characterised the gait of each of the perpetrators. The process was then repeated for each of the suspects. The list of features of gait for each of the perpetrators was then compared to that for each of the suspects, the features considered as being either compatible features, features that were different but were not incompatible, or features that were incompatible (features that would preclude the perpetrator from being the suspect).

Conclusions: Once the comparison was complete, the limitations of the footage were considered and noted. The following limitations were identified:

a. the limited lighting and frame rate of some of the questioned footage
b. the limited resolution and lighting of some of the reference footage
c. the limited frame rate of the reference footage
d. the limited perspectives of the figures/subjects shown by the footage
e. the figure in some of the questioned footage is carrying various objects, which may have affected their gait

In light of the information gained from the comparison of features of gait and taking into account the limitations identified, final conclusions were then reached as to the probative value of the gait analysis evidence for each of the comparisons. These were expressed using the verbal scale described earlier in this book, and described the strength of the evidence to either support or reject the proposition that the perpetrator was the suspect (Figure 13.3).

An independent verifier endorsed the findings, and the report was submitted to the police.

Outcome: Suspect D, with moderate evidence to support the proposition that he was perpetrator 1, was charged with orchestrating the arson attack. Suspect B with moderately strong evidence to support the proposition that he was perpetrator 2, was charged with arson. Suspect B then admitted his guilt, but suspect D did not and the case went to trial where the gait analysis evidence was presented as one facet of the total evidence. Suspect D was found guilty by a jury.

	Perpetrator 1	Perpetrator 2
Suspect A	moderate evidence to reject	moderately strong evidence to reject
Suspect B	limited evidence to reject	moderately strong evidence to support
Suspect C	provides no assistance	limited evidence to support
Suspect D	moderate evidence to support	limited evidence to reject

FIGURE 13.3 Levels of support or rejection for Case Study 2.

Learning points:

1. All of the learning points raised in Case Study 1 can also be applied in this case. However, this multi-comparison case raises several other learning points.
2. Ensure the person in each piece of footage is the same person. When presented with multiple pieces of footage from different cameras, it must be established that the figure is the same person in each piece. This can be done through sequential time stamped videos, or through a suspect agreeing that they are the person in each piece, or through other forensic methods such as body mapping or clothing analysis.
3. Multiple comparisons will take considerably more time. Ensure you factor in the extra time it will take to complete multiple analyses, and to compare them. Provide a realistic quote at the start of the process.
4. Your report must be clear and concise. Although the case may have been complex, the communication of the analysis, comparisons, evaluation, and conclusions must be clear and concise. Impenetrable reports can lead to misunderstandings, may provide the opposition barrister with ammunition to derail your testimony, may reduce the credibility of your testimony, and may undermine you as an expert.
5. With multiple figures it is likely that some features of gait will be seen to be exhibited by more than one of the figures. Furthermore, if some of the figures are related by birth it is potentially more likely that they will demonstrate similar gait characteristics. Firstly, ensure you discuss this with your commissioner as this may be of significance when comparing results, thereby reducing the strength of the probative value. It is a possibility that comparisons of family members may only produce a maximum of limited support, whatever the footage supplied. Secondly, ensure you are methodical and objective at all times. This will inhibit any possible biases when analysing gait, particularly when the gaits are similar and it may be possible for you to recall the gait of the previous subject as you go along.
6. Treat each comparison as a separate case. Avoid seeing each comparison as a part of a whole, and be objective when comparing each analysed gait from figure to subject. Not only will it keep unconscious bias at bay, but it will keep you focussed on one figure/subject at a time.

CASE STUDY 3: TWO ARMED BANK ROBBERIES

Ivan Birch

Background: Two armed robberies occurred at locations some five miles apart, in an urban area of the UK. Both were carried out by four perpetrators using a similar modus operandi and using similar weapons. A forensic gait analyst was contacted by the senior investigating officer of the police team and a request was made for the analysis and comparison of footage captured at the scenes of the crimes showing two of the four perpetrators and of two suspects in the vicinity of the crimes at earlier dates.

Question: Are the two perpetrators the two suspects? As there were two robberies and two sets of reference footage, in response to the question posed by the investigating team, the forensic gait analyst asked two questions in return:

1. Is there evidence that the perpetrators of the two robberies are the same two persons?
2. Is there evidence that the two suspects seen in each of the two sets of reference footage are the same two persons?

The answer from the senior investigating officer was in both cases 'yes'. The senior investigating officer also informed the analyst that based on the evidence of other expert witnesses in body mapping and clothing analysis, perpetrator A was only to be compared to suspect 1 and perpetrator B was only to be compared to suspect 2. The agreed questions were therefore:

1. Is perpetrator A, seen in both sets of questioned footage, suspect 1, seen in both sets of reference footage?
2. Is perpetrator B, seen in both sets of questioned footage, suspect 2, seen in both sets of reference footage?

Materials: The questioned footage consisted of footage taken from three cameras in one bank, and four cameras at the other. The resolution, lighting and frame rate were in all cases suitable for use in forensic gait analysis. However, the field of view of the footage from robbery one showed a relatively small area between the entrance and the counter, while the field of view from robbery two showed a larger open area, but which included an area of sloping floor leading from the entrance.

The reference footage was taken from one camera close to the location of robbery one, showing a long stretch of pavement, and from two cameras close to the location of robbery two, each of which showed a short section of pedestrianised road.

Still images were also provided identifying perpetrators A and B in both robberies, and suspects 1 and 2 in both sets of reference footage.

Analysis and comparison: Analysis of the features of gait shown by the perpetrators in the questioned footage was undertaken first, using the Sheffield Features of Gait Tool, followed by analysis of those shown by the suspects in the reference footage, using the same methodology. Based on the assurances of the senior investigating officer regarding the evidence linking the footage, the observations of features of gait for perpetrator A from the two robberies were amalgamated. The same process was adopted for the observations of features of gait for perpetrator B, and suspects 1 and 2 in the reference footage. The summary of the features that characterised the gait of each of the four figures/subjects was therefore based on a significant number of steps captured on various occasions.

The observed features of gait demonstrated by perpetrator A were then compared to those of suspect 1, and the features demonstrated by perpetrator B with those of suspect 2. The limitations of the footage and its content were then considered, and the following limitations identified:

a) the limited number of consecutive mid gait steps shown by the questioned footage
b) the figures in the questioned footage were turning and/or changing the speed of walking during some of the footage, which may have affected their gait
c) the figures in the questioned footage were carrying objects, which may have affected their gait
d) the figures in the questioned footage were walking up or down a slope in some of the footage, which may have affected their gait

Conclusions: Based on the comparison of the features of gait observed for each of the perpetrators in the questioned footage from both robberies and the features of gait observed for each of the suspects close to the two locations, and taking into account the limitations of the footage, it was concluded that the forensic gait analysis evidence provided support for the proposition that the perpetrator was the suspect to the level shown in Figure 13.4.

The report was completed and verified, and was then sent to the senior investigating officer.

	Perpetrator A	Perpetrator B
Suspect 1	limited support	
Suspect 2		moderate support

FIGURE 13.4 Levels of support or rejection for Case Study 3.

One week later the senior investigating officer contacted the analyst. Having discussed the case with the Crown Prosecution Service, it had been decided that the assumption that the perpetrators of the two robberies were the same individuals was unsound, as was the conclusion that the reference footage from the two locations showed the same two individuals. A request was therefore made to the analyst for the report to be amended with the comparisons related to the two robberies now made separately, with separate conclusions.

The summaries of the features of gait were re-assessed, as was the impact of the limitations on the revised conclusions. The revised conclusions were that the forensic gait analysis evidence provided support for the proposition that the perpetrator was the suspect to the level shown in Figures 13.5.

The revised report was completed and verified, and sent to the senior investigating officer.

Two days later the analyst was contacted by the senior investigating officer, who wanted an explanation as to why different conclusions had been reached, using the same footage, in the revised report. It was explained that the probative value that is assigned to the evidence by the expert is determined by a combination of a consideration of the similarities and differences in the features of gait observed being demonstrated by the perpetrator and the suspect, and the limitations of the footage. The original conclusions were drawn on the basis of a combination of the questioned footage from the two robberies, and the reference footage from the two locations close to the banks. The questioned footage from robbery one showed a very limited number of mid gait steps due to the size of the bank, while the questioned footage from robbery two showed more mid gait steps, but some of those steps being taken on a sloping surface. The combination of the information gained from the two sets of questioned footage yielded a useful amount of information about the gait of the two perpetrators. In the case of the reference footage, the footage from the location close to robbery one showed a good number of consecutive mid gait steps being taken by both suspects, while the reference footage from the location close to robbery two showed fewer mid gait steps even though there was footage from two cameras. The number of mid gait steps seen being taken by each of the perpetrators, and each of the suspects,

Robbery one	Perpetrator A	Perpetrator B
Suspect 1	no assistance in addressing the issue	
Suspect 2		limited support

Robbery two	Perpetrator A	Perpetrator B
Suspect 1	limited support	
Suspect 2		no assistance in addressing the issue

FIGURE 13.5 Revised levels of support or rejection for Case Study 3.

of course varied from footage to footage. The limitations noted also had differential impacts on the conclusions that could be drawn from the footage related to each of the robberies.

Outcome: On disclosure of the complete evidence compiled by the prosecution, the forensic gait analysis being one component of the evidence, the two suspects admitted that they were the two perpetrators.

Learning points:

1. As a forensic gait analyst, always insist on exploring the evidence that links pieces of footage together in the opinion of the commissioner. In doing so you can save yourself a considerable amount of time and additional work, and the commissioner a considerable amount of unnecessary expense.
2. As a matter of course, ask the question regarding the continuity of footage at the time of the initial enquiry.
3. Always keep separate notes of the features of gait observed for every individual piece of footage.
4. If you are supplied with additional footage, additional information, or a revised question after you have submitted your report, you should not alter the original report, but write an addendum which explains what happened, and describes your revised analysis and conclusions.
5. Even though the forensic gait analysis only provided limited support for the relevant propositions, it still contributed to the body of evidence, and assisted in gaining the confessions.

REFERENCE

Brentzel, C. 2018, February 28. Gait closed: How a suspect's walk cracked a cold case robbery. Retrieved November 18, 2018, from https://whnt.com/2018/02/27/gait-closed-how-a-suspects-walk-cracked-a-cold-case-robbery/

Appendix 1: The Sheffield features of gait tool

Appendix 1: The Sheffield features of gait tool

Case: Date:

					Q1	Q2		Summary
	Disc ID							
	Ftge ID							
	Ftge Ref							
	Resolution							
	Lighting							
	Frame Rate							
1	A1	symmetrical gait[1]						
	A2	asymmetrical gait[2]						
	A3	erratic gait						
		symmetry of gait could not be determined						
2	B1	no significant rolling of the head and torso[3]						
	B2	rolling of the head and torso						
	B3	rolling of the head and torso, more to the right on right steps than to the left on left steps						
	B4	rolling of the head and torso, more to the left on left steps than to the right on right steps						
		frontal plane motion of the head and torso could not be determined[4]						
3	C1	no significant yawing of the torso[5]						
	C2	yawing of the torso, right side forwards on left steps, left side forwards on right steps						
	C3	yawing of the torso, right side forwards on right steps, left side forwards on left steps						
		transverse plane motion of the torso could not be determined[6]						
	C4	noticeable vertical movement of the head and torso on each step						

[1] A symmetrical gait is one in which the gross movement of the head and torso to the right is approximately the same as the gross movement to the left.
[2] An asymmetrical gait is one in which the gross movement of the head and torso to the right is different to the gross movement to the left.
[3] Rolling of the head and torso is a rotation that would be best seen from the front or back.
[4] The frontal plane is the plane of body movements which are best seen from the front or rear of the body.
[5] Yawing of the torso is a rotation that would be best seen from above.
[6] The transverse plane is the plane of body movements which are best seen from above.

Appendix 1: The Sheffield features of gait tool

4	D1	head held approximately in line with the midline of the torso in the frontal plane			
	D2	head held tilted to the right relative to the midline of the torso in the frontal plane			
	D3	head held tilted to the left relative to the midline of the torso in the frontal plane			
		frontal plane alignment of the head could not be determined			
5	E1	head held approximately in line with the midline of the torso in the sagittal plane[7]			
	E2	head held forwards of the midline of the torso in the sagittal plane			
	E3	head held tilted forwards relative to the midline of the torso in the sagittal plane			
	E4	head held tilted backwards relative to the midline of the torso in the sagittal plane			
		sagittal plane alignment of the head could not be determined			
6	F1	shoulders relatively level			
	F2	right shoulder lower than the left shoulder			
	F3	left shoulder lower than the right shoulder			
	F4	dipping of the right shoulder on right steps more than the left shoulder on left steps			
	F5	dipping of the left shoulder on left steps more than the right shoulder on right steps			
		relative height of the shoulders could not be determined			
7	G1	magnitude of right arm swing limited			
	G2	magnitude of left arm swing limited			
	G3	magnitude of right arm swing unremarkable			
	G4	magnitude of left arm swing unremarkable			
	G5	magnitude of right arm swing substantial			
	G6	magnitude of left arm swing substantial			
	G7	magnitude of right arm swing erratic			
	G8	magnitude of left arm swing erratic			
		magnitude of arm swing could not be determined			

[7] The sagittal plane is the plane of body movements which are best seen from either side of the body.

8	H1	path of motion of the right hand approximately parallel to the line of progression
	H2	path of motion of the left hand approximately parallel to the line of progression
	H3	path of motion of the right hand oblique, the hand more medial anteriorly than posteriorly
	H4	path of motion of the left hand oblique, the hand more medial anteriorly than posteriorly
	H5	path of motion of the right hand erratic
	H6	path of motion of the left hand erratic
		orientation of the path of motion of the hand could not be determined
9	I1	more right arm swing movement anterior of the torso than posterior of the torso
	I2	more left arm swing movement anterior of the torso than posterior of the torso
	I3	right arm swing movement anterior and posterior of the torso approximately equal
	I4	left arm swing movement anterior and posterior of the torso approximately equal
	I5	comparative anterior/posterior movement of right arm swing erratic
	I6	comparative anterior/posterior movement of left arm swing erratic
		comparative anterior/posterior arm swing movement could not be determined
10	J1	right elbow is fully or close to fully extended throughout arm swing
	J2	left elbow is fully or close to fully extended throughout arm swing
	J3	right elbow is flexed throughout arm swing
	J4	left elbow is flexed throughout arm swing
	J5	flexion at the right elbow increases during the forward motion of the arm
	J6	flexion at the left elbow increases during the forward motion of the arm
		angulation at the elbow could not be determined

11	K1	largely linear motion occurring at right hip[8]
	K2	largely linear motion occurring at left hip
	K3	largely linear motion occurring at the right hip, with slight circumduction occurring prior to heel strike[9]
	K4	largely linear motion occurring at the left hip, with slight circumduction occurring prior to heel strike
	K5	circumduction occurring at right hip
	K6	circumduction occurring at left hip
		nature of the movement occurring at the hips could not be determined
12	L1	right knee points inwards, towards the midline of the body, at feet adjacent during swing
	L2	left knee points inwards, towards the midline of the body, at feet adjacent during swing
	L3	right knee points forwards approximately in line with the direction of travel at feet adjacent during swing
	L4	left knee points forwards approximately in line with the direction of travel at feet adjacent during swing
	L5	right knee points forwards, or slightly inwards/outwards, at feet adjacent during swing
	L6	left knee points forwards, or slightly inwards/outwards, at feet adjacent during swing
	L7	right knee points outwards, away from the midline of the body, at feet adjacent during swing
	L8	left knee points outwards, away from the midline of the body, at feet adjacent during swing
	L9	right more so than left
	L10	left more so than right
		the comparative orientation of the knees at feet adjacent during the swing phase of gait could not be determined
		orientation of the knees at feet adjacent during the swing phase of gait could not be determined
13	M1	right knee is flexed at or close to heel strike
	M2	left knee is flexed at or close to heel strike
	M3	right knee reaches a position of full or close to full extension at or close to heel strike
	M4	left knee reaches a position of full or close to full extension at or close to heel strike
	M5	right knee reaches a position of full or hyper extension at or close to heel strike[10]
	M6	left knee reaches a position of full or hyper extension at or close to heel strike
		angulation at the knees at or close to heel strike could not be determined

[8] Linear movement in this instance is a movement occurring at the hip joint that results in the lower limb moving forward in an approximately straight line during the swing phase of gait.
[9] Circumduction in this instance is a movement occurring at the hip joint that results in the lower limb moving forward in an arc during the swing phase of gait.
[10] Full extension of the knee is a position in which the thigh and shank are aligned. Movement beyond this point is known as hyperextension, the knee appearing to fold backwards.

Appendix 1: The Sheffield features of gait tool

14	N1	right knee is flexed prior to heel rise								
	N2	left knee is flexed prior to heel rise								
	N3	right knee reaches a position of full or close to full extension prior to heel rise								
	N4	left knee reaches a position of full or close to full extension prior to heel rise								
	N5	right knee reaches a position of full or hyper extension prior to heel rise								
	N6	left knee reaches a position of full or hyper extension prior to heel rise								
		angulation at the knees prior to heel rise could not be determined								
15	O1	base of gait is very narrow[11]								
	O2	base of gait is narrow								
	O3	base of gait is moderate								
	O4	base of gait is wide								
	O5	base of gait is variable								
		base of gait could not be determined								
16	P1	right foot points inwards, towards the midline of the body, when weight bearing during stance								
	P2	left foot points inwards, towards the midline of the body, when weight bearing during stance								
	P3	right foot points forwards, approximately in line with the direction of travel, when weight bearing during stance								
	P4	left foot points forwards, approximately in line with the direction of travel, when weight bearing during stance								
	P5	right foot points forwards, or slightly inwards/outwards, when weight bearing during stance								
	P6	left foot points forwards, or slightly inwards/outwards, when weight bearing during stance								
	P7	right foot points outwards, away from the midline of the body, when weight bearing during stance								
	P8	left foot points outwards, away from the midline of the body, when weight bearing during stance								
	P9	right more so than left								
	P10	left more so than right								
		the comparative orientation of the feet when weight bearing during stance could not be determined								
		orientation of the feet when weight bearing during stance could not be determined								

[11] The base of gait is the side to side distance between the heels on consecutive steps.

Appendix 1: The Sheffield features of gait tool

17	Q1	right foot points inwards, towards the midline of the body, at feet adjacent during swing
	Q2	left foot points inwards, towards the midline of the body, at feet adjacent during swing
	Q3	right foot points forwards, approximately in line with the direction of travel, at feet adjacent during swing
	Q4	left foot points forwards, approximately in line with the direction of travel, at feet adjacent during swing
	Q5	right foot points forwards, or slightly inwards/outwards, at feet adjacent during swing
	Q6	left foot points forwards, or slightly inwards/outwards, at feet adjacent during swing
	Q7	right foot points outwards, away from the midline of the body, at feet adjacent during swing
	Q8	left foot points outwards, away from the midline of the body, at feet adjacent during swing
	Q9	right more so than left
	Q10	left more so than right
		the comparative orientation of the feet at feet adjacent during swing could not be determined
		orientation of the feet at feet adjacent during swing could not be determined
18	R1	right foot is noticeably inverted prior to heel strike[12]
	R2	left foot is noticeably inverted prior to heel strike
	R3	right foot noticeably everts on weight bearing
	R4	left foot noticeably everts on weight bearing
19	S1	right forefoot is markedly raised prior to heel strike
	S2	left forefoot is markedly raised prior to heel strike
	S3	right heel raise is relatively early
	S4	left heel raise is relatively early
20	T1	right foot is noticeably abducted prior to heel strike[13]
	T2	left foot is noticeably abducted prior to heel strike
21	U1	during the early part of the stance phase of gait, the thigh of both lower limbs appears slightly inverted[14]

[12] An inverted foot describes an orientation of the foot in which the sole is turned towards the midline of the body when the foot is viewed from the front or back.
[13] An abducted foot describes an orientation of the foot in which the toes are further away from the midline of the body than the heel when the foot is viewed from above or below. In the case of this feature of gait (for the right and left foot), this orientation is achieved after feet adjacent and before heel strike.
[14] An inverted thigh describes an orientation of the thigh in which the lower end of the thigh is closer to the midline of the body than the upper end of the thigh.

Index

ACE-V, *see* Analysis, Comparison, Evaluation, and Verification
AFSP, *see* Association of Forensic Science Providers
Analysis, Comparison, Evaluation, and Verification (ACE-V), 35, 90
Analyst, process for, 89–90
Ankle
    feet adjacent and, 47
    heel rise and, 45–46
    initial foot contact and, 43
    loading response and, 43–44
    mid-stance phase and, 45
    opposite initial contact and, 46
    opposite toe off and, 44
    tibia vertical and, 48
    toe off and, 47
Appendices, in report writing, 139–140, 146
Arm swing, 110
Association of Forensic Science Providers (AFSP), 172–174

BC Court of Appeal, 63
Bertillon, Alphonse, 20
Bertillonage, *see* Signaletics
Bias
    cognitive, 89–90, 93, 121, 162, 164–165, 197
        strategies to minimize, 165–166
    confirmation, 164
    mplicit/unconscious bias, 165
Biometrics, 11–13, 13n3
British Columbia Court of Appeal, 27

Cadence, 22, 49, 97, 98
Camera setup, for video analysis of gait, 55
Canada, 4, 27, 60–65

Carrying, of objects, 111
CCTV, *see* Video footage
Chartered Society of Forensic Sciences, 73, 76, 80, 81, 87
    Certificate of Professional Competence, 73–74, 79
    Forensic Gait Analysis Working Group, 76, 82
    Standards Committee, 73
Checklist, of gait analysis, 51–52
Civil proceedings, gait evidence in, 33–34
College of Podiatry, 8, 14n2, 76, 81
    Forensic Podiatry Special Advisory Group, 81
    Forensic Podiatry Special Interest Group, 75
Commissioning, of forensic gait analysis, 87–89
Commonwealth v. Caruso (2014), 25
Commonwealth v. Kinney (2019), 32, 33
Commonwealth v. Spencer (1994), 23
Compatible features and gait features, 118
Connecticut Court of Appeals, 29
Corroborative footage, 98
Council for the Registration of Forensic Practitioners, 72, 74
Court of Appeals (US), 23–25
Criminal Code (Canada), Section 657.3, 63
Criminal Practice Directions (UK), 67, 77
Criminal Procedure Rules (UK), 67, 77, 82
Criticisms and challenges, 8–10
Cycle time, 42, 50

Daubert criteria, 7, 8, 62
Daubert factors, 7
Daubert v. Merrell Dow Pharmaceuticals Inc. (1993), 27, 61, 63

213

Index

Declarations, in report writing, 135, 146
Double limb support, 40–42
Double support stance, 40

Edinburgh Visual Gait Score, 51
ENFSI, *see* European Network of Forensic Science Institutes
EU General Data Protection Regulation, 82
Europe, 4, 123, 172, 181, 185
European Network of Forensic Science Institutes (ENFSI), 172–174, 177
scale, 123
Evaluative opinion, 173–174
   in gait analysis, 174–175
      assignment of probabilities, 175–176
      propositions, 175
      relevant circumstances, 175
Evaluative report, 103n3
Evidence presentation, in court, 147–148
   arriving at court for, 150, 152
   giving, 153–156
   with notes and report, 148–149
   post-, 157
   and preparation for trial, 150
   pre-trial meeting, 149, 151
   requirement notification for, 149
   site visits for, 151
   still images as, 156–157
Evidential reporting, significance of, 122
Exhibits
   examination of, 145
   list of, 144
Expert opinion evidence
   in Canada
      admissibility, 63–64
      reliability and scientific validity, 63
   criteria for, 28
   trial judge and, 61–63
   in United States, 60–62
Expert witness, ethical role and responsibilities, 64–65
Expert witness reports, writing, 129–130
   checklist for, 141–142
   forensic gait analysis reports and, 144
      analysis and comparisons, 145
      appendices, 146
      declarations, 146
      evaluations and conclusions, 146
      exhibits, 145
      name and relevant qualifications and experience, 144
      task requested and undertaken, 144
   general contents of, 134–135
      appendices, 139–141
      background information, 136
      conclusion and evaluation, 138–139
      declarations, 135
      discussion, 138
      instruction, 136
      introduction, 135–136
      methods and analytical techniques, 137
      references, 139
      results, 137–138
      samples supplied, 136–137
   key principles of, 130–134
   legal requirements of, 140–141
   potential pitfalls in, 141–142
   verification and submission and, 141
Eyewitness testimony, 24, 31, 32

Family and Medical Leave Act (FMLA), 33
Federal Rules of Criminal Procedure, 61
Federal Rules of Evidence
   Rule 701 (US), 23
   Rule 702, 61, 62
Feet adjacent, 45, 47
Figure and subject, differential use of, 95–96
Fingerprint matching, significance of, 162–163
Flesch–Kincaid (FK) grade level, 131
Flight phase, 43
FMLA, *see* Family and Medical Leave Act
Footwear/lack of, 110
Forensic entomology, 81
Forensic gait analysis, legal context of
   evaluative phase, 60
   investigative phase, 59–60
   in North America, 59–60
      expert witness ethical role and responsibilities, 64–65
      law principles and legal system, 60–64
   in United Kingdom, 67–69
Forensic Gait Analysis: A Primer for Courts (Royal Society and the Royal Society of Edinburgh), 7, 9, 78, 124, 172
Forensic gait analysis, definition, 2
Forensic gait analysis reports, writing, 144–146
Forensic Podiatry Role and Scope of Practice (International Association for Identification document), 73, 75
Forensic Science in Criminal Courts (PCAST report), 63
Forensic Science Regulator (UK), 8, 13n2, 25, 73–77, 80–82
   Codes of Practice and Conduct for Forensic Science Providers and Practitioners in the Criminal Justice System, The, 8, 68, 74, 76, 78, 80, 81, 107
   Code of practice for forensic gait analysis, 3, 8, 25, 69, 78, 81–82, 90, 125
   creation of, 72

on declarations, 135
Expert Report Guidance, 140
Forensic Science Society, *see* Chartered Society of Forensic Sciences
Forensic unit, 78
Frame rate, of video footage, 96–98

Gait
  comparison, 117–121
    evaluation and, 121–124
    payment and, 127
    verification and, 125–127
  cycle
    in detail, 43–50
    events of, 41–42
    illustration of, 41
    phases of, 41–42
    single and double support timing in, 40–41
  out-toed, 189–190
  speed of, 48–49
  and walking, comparison of, 1
Gait evidence
  admissibility, 27–29, 63
  types, 34
General Electric Co v Joiner (1997), 62
Gibson v. Texas (2019), 32
Google Maps, 110, 151
Governance, for forensic gait analysis practice, 8

Handcuffs, 110–111
Hard data, 175–176, 179, 180
Health and Care Professions Council, 82
Heel rise, 45–46
Heel strike, *see* initial foot contact
Heuristics, *see* Perceptual error psychology
Hip
  feet adjacent and, 47
  heel rise and, 45–46
  initial foot contact and, 43
  loading response and, 43–44
  mid-stance phase and, 45
  opposite initial contact and, 46
  opposite toe off and, 45
  tibia vertical and, 48
  toe off and, 47

Identification from gait, significance of, 6–7, 9
Impartiality and professional objectivity, for expert report, 132–133
In-app technology, 54
Inattentional blindness, 160–162
Incompatible features and gait features, 118–119
Initial contact, with potential commissioner, 90–93

Initial foot contact, 43–44
International Association for Identification, 64, 73, 75
International Laboratory Accreditation Cooperation, 81
Investigative opinion, 172–174
Investigative report, 103n2

Knee
  feet adjacent and, 47
  heel rise and, 45–46
  initial foot contact and, 43
  loading response and, 43–44
  mid-stance phase and, 45
  opposite initial contact and, 46
  opposite toe off and, 44
  tibia vertical and, 48
  toe off and, 47
Kumho Tire Co v Carmichael (1999), 62

Language and style of writing, for expert report, 130–132
Layout, of expert report, 133
Leading leg, 41
Likelihood ratio, 176, 178–181, 185
List approach, to gait features observation, 119–120
Loading response phase, 43–44

McGurk effect, 160
  heuristic error in, 162
Massachusetts Court of Appeals, 33
Michigan v. Ballard (2016), 31, 35
Mid-stance phase, 45
Mid-stance rocker, 45

National Crime Agency (UK), 87
Non-evidential reports, 130
Notes in forensic work, significance of, 107–111
Nyquist critical frequency, 97

Observational gait analysis
  in clinical and forensic contexts, 89
  from video footage, 10–11
  without video recording, 50–52
Opposite initial contact, 46
Opposite toe off, 44
Optoelectronic stereophotogrammetry, 55
Oregon Court of Appeals, 30

People v. Beasley (2018), 32
People v. Pratt (2019), 33
Perceptual error psychology, 159
  focus on forensic science and, 162–164
  and forensic gait analysis
    potential pitfalls, 164–165

strategies to minimize bias, 165–166
human perception limits and, 160–162
Persistence of vision, 96
Phi phenomenon, 96
Podiatry, 5, 22, 25–29, 33, 72–76
Probative value, 13, 13n4, 98, 102, 103n3, 122–123, 171–172, 199, 203
   AFSP and ENSI inferential paradigm and, 172–174
   evaluative opinion and, 174–176
   expression of, 123
   gait observations evaluation versions and, 176–185
   verbal expressions scale of, 123–124, 199
Pronation, of foot, 56n1
Proximity, of figures, 111

Quality framework and forensic environment, relationship between, 77

Readability, of expert report, 132–134
   ways to improve, 131–132
Readability statistics, 131
Report writing, *see* Expert witness reports, writing
R. v. Aitken (2008), 27, 28, 63, 64
R. v. J.-L.J. (2000), 27, 63
R. v. Mohan (1994), 28
R. v. Otway (2011), 28, 29
R. v. Saunders (2000), 25, 35
R v Sekhon (2014), 63
R v Trochym (2007), 64

Scientific knowledge, 61
   and validity, considerations of, 62
Screening report, 103n2
Sense-making, 161
Sensitivity analysis, significance of, 176
SFR1, *see* Streamlined Forensic Reports
Sheffield Features of Gait Tool, 7, 31, 108, 113, 120, 180, 190, 197, 202, 204–211
Signaletics, 20
Single limb support, 40–42
Single support stance, 40
Skills for Justice, 74
Slope, of ground, 110
Society of Chiropodists and Podiatrists, *see* College of Podiatry
Soft data, 175, 176, 179, 180
Software, purpose-designed, 10
Spreadsheet approach, to gait features observation, 120
Stance phase, 40
Stance time, 42
State v. Coyne (2010), 29

State v. Fivecoats (2012), 29, 30
State v. Iverson (1971), 21, 22
State v. Mark (1980), 22–23, 34
State v. Santos (2007), 24
State v. Williams (1996), 24, 35
Steps
   fully seen, 109
   mid gait, 103n1, 109–110, 203
   non-mid gait, 110
   partially seen, 109
Still images, as evidence in court, 156–157
Streamlined Forensic Reports (SFR1), 130
Strengthening forensic science in the United States (US NRC report), 62, 72, 73, 130
Strickland v. Washington (1984), 35
Stride length, 49–50
Supination, of foot, 56n2
Supreme Court of Canada, 63
Swing phase, 40–42
Syms v. Warden (2003), 26

Temporal and spatial parameters of gait, 48, 49
Terminal foot contact, 48
Terminal rocker phase, 46
Three-dimensional motion analysis, 55
Three-dimensional video analysis, 54
Tibia, 43–46
Tibia vertical, 48
Tillman v. Ohio Bell Tel. Co. (2011), 33
Time lapse footage, 97
Toe off, 40–42, 44–47, 109
Trailing leg, 41, 44
Trial judge, 7, 26–28, 30, 61–64
Trunk
   tibia vertical and, 48
   toe off and, 47

UK's House of Lords Science and Technology Committee, 163
UK Law Commission, 74
United Kingdom Accreditation Service, 74, 77, 80
United Kingdom House of Commons Science and Technology Committee report, 72
United States ex rel. Bass v. Ahitow (1998), 24
United States National Academy of Sciences, 163
University of Staffordshire, 73
Upper body
   feet adjacent and, 47
   heel rise and, 45–46
   initial foot contact and, 43
   loading response and, 43–44

mid-stance phase and, 45
opposite initial contact and, 46
opposite toe off and, 44
US Organization of Scientific Area Committees for the Forensic Sciences, 64
US Supreme Court, 62

Verbal expressions scale, of probative value, 123–124, 199
Video footage, 6, 93
-based gait analysis, 52–55
CCTV, 28–29
cognitive bias in, 165–166
cost effectiveness and implications of, 9–10, 100
frame rate of, 96–98
initial contact with potential commissioner for, 90–93
measurements from, 11
observational gait analysis from, 10–11
preliminary assessment of, 93–100
questioned, 10, 88, 90, 92–94, 96, 99–101, 113, 114, 120–122, 126, 156, 164, 177–179, 181–185, 190–199, 201–203
reference, 10, 88, 90, 92–94, 96, 99–102, 114, 120, 122, 126, 164, 177–179, 182–184, 190–209, 201–203
selection criteria of, 88
software, significance of, 94
still images as evidence from, 156–157
superimposed, 120–121
surveillance, 23, 25
task definition with, 101–103
tool for quality assessment of, 94–95
variations in observations of, 112–113
Vision, significance of, 159–161

White Burgess Langille Inman v Abbott & Haliburton, 63, 65
Witness box, evidence presentation in, 155

Zoopraxiscope, 52

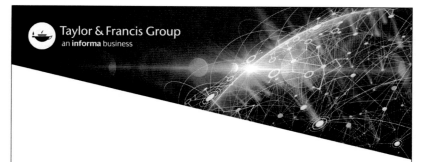